Christian SEKIMONYO SHAMAVU
Boaz BANDU BALUME

Biogas production

Christian SEKIMONYO SHAMAVU
Boaz BANDU BALUME

Biogas production

Feasibility study

ScienciaScripts

Imprint

Any brand names and product names mentioned in this book are subject to trademark, brand or patent protection and are trademarks or registered trademarks of their respective holders. The use of brand names, product names, common names, trade names, product descriptions etc. even without a particular marking in this work is in no way to be construed to mean that such names may be regarded as unrestricted in respect of trademark and brand protection legislation and could thus be used by anyone.

Cover image: www.ingimage.com

This book is a translation from the original published under ISBN 978-620-6-69621-6.

Publisher:
Sciencia Scripts
is a trademark of
Dodo Books Indian Ocean Ltd. and OmniScriptum S.R.L publishing group

120 High Road, East Finchley, London, N2 9ED, United Kingdom
Str. Armeneasca 28/1, office 1, Chisinau MD-2012, Republic of Moldova, Europe
Printed at: see last page
ISBN: 978-620-7-23709-8

Copyright © Christian SEKIMONYO SHAMAVU, Boaz BANDU BALUME
Copyright © 2024 Dodo Books Indian Ocean Ltd. and OmniScriptum S.R.L publishing group

Contents

Epigraph	2
GENERAL INTRODUCTION	3
Chapter I	6
Chapter II	14
Chapter III	30
Chapter IV	42
GENERAL CONCLUSION	49
Bibliography	51
Appendix	55

Epigraph

"The planet is a common good that we have a duty to preserve now and for future generations. In the face of this immense challenge, renewable gases are part of the solution."

AFG (2018)

Boaz BANDU BALUME

GENERAL INTRODUCTION

1. State of the question

Biogas is a clean, renewable source of energy produced from the decomposition of organic matter such as food waste, agricultural waste and sewage sludge (Belanger, 2009; p. 39).

Its production is a global activity that is attracting growing interest in the fields of energy, development and the environment, due to its interdisciplinary nature (Laura, 2019; p:4).

Thus, many authors have investigated this topic in different regions of the world. Some of these predecessors, such as Wendy (2017, p: 32), have studied the limitations of biogas digesters in flexible operation, focusing on the impact of load variations on the anaerobic digestion process and on biogas production.

However, RECORD (2013; p. 151) points out that: "*the production and recovery of biogas from different substrates of agricultural, urban and industrial origin are at the heart of sustainable development issues, enabling the production of renewable energy, the reduction of greenhouse gases and the treatment of waste*".

Hence Belanger (2009; p: 1) [5], specifies that biogas is one of the oldest sources of energy considered to be green. It is essentially composed of methane (CH_4), carbon dioxide (CO_2), a variable amount of water and hydrogen sulphide in different proportions depending on the substrate used (Fernandez, 2021; p:29).

In addition, Thomas J. et al (2020; p. 23) point to a number of problems associated with energy production and use in Africa, such as excessive dependence on fossil fuel sources, poor access to energy in some regions, lack of infrastructure for energy production and distribution, and the negative environmental consequences associated with the use of fossil fuels.

In such a situation, the use of green energies, such as biogas, seems to be a good alternative (Bautista, 2019; p. 20); because, methanisation is a technology that can help reduce greenhouse gas emissions, as highlighted by Bautista (2019;p. 80) in his doctoral thesis on "Feasibility study of micro-methanisation by co-digestion at neighbourhood scale" and supported by Cedric Philbert (2018), who proposes the co-digestion of different organic matter and the implementation of efficient waste treatment systems.

Although the Record report (2022; p. 118) highlights the technical and economic challenges associated with introducing digestate to maximise biogas production, technologies exist to optimise methanisation, and pretreatments have been proven to improve performance (Vincent, 2013; p. 26).

To facilitate the adoption and use of biogas, Zheng, Xu et al. (2015; p. 254) also explored the advantages and disadvantages of biogas and proposed solutions to improve its use.

Furthermore, in many cities around the world, urban waste and water hyacinth are often poorly managed and sources of pollution, whereas they could be exploited to produce renewable energy, as noted by Lardon and Ribeiro (2012; p. 13) in their book entitled "biomethanisation" and Njogu, Kinyua et al, (2015; p. 22) in their book "Biogas production using water hyacinth (Eicchornia crassipes) for power generation in Kenya".

In urban areas such as Goma, where there is strong population growth, biogas production could prove to be an innovative solution for meeting energy needs while promoting sustainable waste management and a healthy environment.

The production of biogas could therefore offer the city of Goma an opportunity to access electricity, as Almalowi and Jobran (2022; p. 7) have experimented by converting biodegradable waste into biogas.

However, the design and implementation of digesters at local level can present certain difficulties. This is why we are looking at small-scale production as a viable alternative for reducing greenhouse gases.

2. Choice and interests of the subject
2.1. Choice of subject

The choice of this subject is highly relevant because it addresses a major problem for cities in developing countries, and especially the city of Goma, where access to a reliable source of energy is often limited and where waste management is a real challenge. That's why we chose this topic.

2.2. Interests of the subject

For us personally, this research will enable us to discover new ways of producing biogas, and could create new jobs for young people and workers, thereby improving economic and social conditions in Goma.

Biogas can reduce dependence on fossil fuels, which are responsible for greenhouse gas emissions, and can also help to manage the city's biodegradable waste, thereby preserving the environment. Our feasibility study on biogas production will be a valuable guide for future scientific research and improvements in this field.

3. Problems

The growing demand for energy around the world goes hand in hand with a dynamic population whose demographic growth is incredibly high. Our society is constantly evolving and positioning itself in relation to the world around it. This evolution brings with it a number of reflections, particularly on the relationship between our society and the planet's natural resources.

In this situation, the use of green energies, including biogas, seems interesting and beneficial for complementarity (Bautista, 2019; p. 78). Biogas is one of the oldest sources of energy considered to be green. It is produced by the anaerobic decomposition of biomass and is mainly composed of methane (CH_4) (50 to 85%), carbon dioxide (CO_2) and varying amounts of water and hydrogen sulphide, depending on the substrate used (Belanger, 2009; p. 34).

In general, biogas production has many advantages for developing countries, and can be an interesting solution to meet their energy needs while respecting the environment and contributing to economic and social development (Kretschmer et al., 2012). However, there is a need to study the technical, environmental and social issues associated with the implementation of this technology, and to put in place clear policies and financial support to encourage its adoption and dissemination in these developing countries.

4. Research questions

In order to carry out our study properly, we will mainly answer the following question:

A. Is it possible to produce biogas in Goma?

Specifically, this question raises the following issues:

B. What are the most important sources of organic matter in Goma for producing biogas?
C. What obstacles are there to biogas production in Goma?
D. What strategies are needed to set up an efficient biogas production infrastructure in the city of Goma?

5. Working hypotheses
5.1. Main hypothesis

The biogas production process could be possible in the town of Goma using an anaerobic digestion plant that converts organic waste into biogas and digestate.

5.2. Specific assumptions

A. The resources available to produce biogas in the city of Goma would be organic waste of animal and plant origin, which could be collected and used as raw materials for biogas production. It would also be possible to grow specific energy crops to produce lignocellulosic materials for biogas production.

B. Insufficient information and material, financial and technical or technological resources would be one of the major difficulties encountered in producing biogas in the city of Goma.

C. An awareness and knowledge campaign on biogas; the use of simple technology with locally accessible equipment; ensuring the regular collection of organic matter and adequate access to storage and distribution infrastructure to facilitate the consumption of biogas in the city of Goma would be the strategies to be adapted to achieve biogas production in the city of Goma.

6. Objectives

6.1 Main objective

The main aim of this study is to examine the feasibility of producing biogas in the town of Goma.

6.2 Specific objectives

However, we have set ourselves the following specific objectives:

1. Identify the most important sources in Goma for producing biogas.
2. List the difficulties that may be encountered in implementing biogas production in Goma.
3. Propose strategies that can be adopted to produce biogas in the town of Goma.

4. Spatial and temporal delimitation

Our study will be carried out in the town of Goma in North Kivu in the Democratic Republic of Congo from April to August 2023.

5. Labour division

Apart from the introduction and conclusion, our work will be divided into four general chapters, which are presented as follows:
- Chapter I will deal with the LITERATURE REVIEW,
- Chapter II will deal with RESEARCH METHODOLOGY,
- Chapter III will PRESENT THE RESULTS, and
- Chapter IV will give a STRATEGY PROPOSAL.

Chapter I
LITERATURE REVIEW
The first chapter of this study explores various key concepts relating to the production of biogas from organic matter. Through a literature review, we dive into an in-depth exploration of the fundamental definitions, theoretical approach and current knowledge surrounding organic materials, biogas production and biogas itself.

1.1. Definition of concepts
1.1.1. Production
According to Samuelson P. (2009, p. 79), production is the process of creating and transforming resources into goods and services to satisfy the needs and wants of consumers. He goes on to say that "production involves the use of factors of production such as labour, capital and raw materials to produce goods and services that contribute to economic growth and human well-being".

1.1.2. Biogas
According to Sekimonyo Shamavu Christian (2023; p.14) in his book "problematiques energetiques: transition energetique", biogas is a gas most often produced by the fermentation of organic elements under the action of micro-organisms. This fermentation process, which occurs spontaneously in natural ecosystems (marshes, rice paddies, soils, mammal intestines, etc.), is called methanisation (Sekimonyo, 2023; p15). It has now been mastered by mankind and enables us to recover our waste while at the same time replacing certain fossil energy sources. The result: a reduction in our greenhouse gas emissions (https://www.ifpenergiesnouvelles.fr; consulted on 13 April 2023 at 00:03:00).

1.1.3. A Study
The version (2006; p.27) of the Larousse dictionary indicates that study is a systematic process of research, exploration and learning about a particular subject in order to develop knowledge and a deeper understanding.

In addition, a social science study is a specialised examination of the human being, his culture and his relationship with his environment (Kitaganya and Sekimonyo, 2023; p. 23); hence man as transcendence and as the presence of the infinite (Kitaganya, 2023; p. 17).

Thus; Professor KAMPEMPE BUSILI Justin DUPAR (2014; p. 46) defines a study in the exact sciences as one that lends itself to precise and reproducible calculations; the corollary being that they do not depend on the appreciation of the observer. Hence, in the exact sciences, the term 'study' can be understood as 'an analysis aimed at answering any question with measurable and reproducible precision; and can be expressed analytically and objectively' (Bukanga and Sekimonyo, 2021; p. 61); hence it is a study which, for its methodological rigour (scientific method), is capable of producing results and predictions with a precise quantitative expression (Kampempe and Sekimonyo, 2022; p.70).

Finally, Kitaganya and Sekimonyo (2023; p. 32) point out that in the applied sciences, a study is an experimental analysis undertaken in order to acquire new practical knowledge.

1.1.4. Feasibility
The definition of "feasibility" may vary slightly. However, in general, feasibility refers to the technical, economic and organisational possibility of achieving a given project, action or objective (Jacques Gregoire and Andre Petitjean, 2006, p. 22).

Some authors, such as J-P. Bonet (2011), place greater emphasis on technical feasibility, which assesses whether the project can be technically feasible using the resources, skills and knowledge available.

Finally, organisational feasibility assesses whether the project can be implemented effectively and successfully, taking into account factors relating to human resources, organisational structure, processes and the necessary coordination (Harold Kerzner, 2017).

It is important to note that the precise definition of feasibility may vary depending on the specific context and the publications of different authors.

1.1.5. Feasibility study

A feasibility study is an analysis aimed at verifying the viability and practical consequences of the object of study (Matamba and Sekimonyo. 2022; p. 93); therefore, a feasibility study can be carried out as part of an innovation, verification, confirmation, invalidation or experiment (Matundu and Sekimonyo. 2021; p. 11).

1.2. Theoretical approach

Biogas, being a gas produced by the degradation of organic matter in the absence of oxygen (Wendy LAPERRIERE., 2017, p. 13), is a renewable gas (Francois P, 2010, p. 29), mainly composed of methane (Bernard O, 1987, p. 4). It can come from the degradation of organic matter stored in *non-hazardous waste storage facilities* (ISDND) (Atee Club, 2014; p. 3) or be produced by the methanisation of non-woody organic matter in a digester (RECORD, 2002, p. 130).

Biogas is also produced from biomass (Filde A., 2013, p. 9), and its production also contributes to waste management by reducing the amount of organic matter that ends up in landfill sites (Huong N. at all., 2010, p. 47).

So, to find out a little more, we need to expand on the following literature;

❖ Organic materials
❖ Biogas, and
❖ Biogas production

1.2.1. Organic materials

Organic matter refers to any substance derived from living organisms or their waste (Raymond R Wei, 2016; p. 18). They consist mainly of carbon compounds and also contain elements such as hydrogen, oxygen, nitrogen, phosphorus and sulphur (P. Hardouin, 2018; p. 6).

The organic materials required to feed a biodigester come from various sources such as food and agricultural waste, animal waste, manure, by-products of the agri-food industry, crop residues, etc. (CNR, 2000; p. 5). According to Julien L. at al. (2011), organic waste sources can vary in terms of availability and quality, which can pose challenges for the consistent supply of suitable materials to the biodigester.

According to Akunna et al. (2013), the availability of organic waste can be influenced by factors such as population density, local agricultural practices, eating habits, environmental regulations, etc. The quality of the organic matter, on the other hand, is a key factor in ensuring efficient digestion and optimum biogas production (J. Hess, 2007; p. 201). The quality of the organic matter is an essential factor in ensuring efficient digestion and optimum biogas production (J. Hess, 2007; p. 201). Organic matter containing undesirable substances, such as chemical contaminants or heavy metals, can compromise the biodigestion process and the quality of the biogas produced (Bardi et al., 2017, p. 17).

Organic matter is an important resource in many industrial processes and plays a key role in sustainable development. According to Whelan et al (2009, p. 34), the production of energy from biomass, also known as bioenergy, is based on the use of organic materials. These organic materials are converted into biogas, bioethanol, biodiesel or solid fuels to generate

electricity, heat or fuel (Hamza Lamya at al., 2015; p. 3).
In the field of organic chemistry, organic materials are used as raw materials for the manufacture of chemical products such as plastics, textile fibres, pharmaceuticals, pesticides, etc. (Gelinas, 2013, p. 93). Smith and March (2007, p. 17) explain that the organic carbon present in organic matter serves as the basis for the synthesis of many chemicals and materials used in various industrial sectors.
Recycling organic waste is also a common practice in waste management (Augris, 2002, p. 132). And the CCE (2017, p. 22) adds that this waste can be turned into fertiliser or used in other industrial processes. According to Boldrin et al (2011), recycling organic waste reduces the amount of waste sent to landfill and helps to preserve natural resources.

Categories of Organic materials

Organic matter, also known as organic waste, can be divided into several categories. They are generally classified according to their origin and specific characteristics (Augris, 2000; p. 9). Here are some of the main categories of organic matter:

- **Food waste:** This is food leftovers from households, restaurants, supermarkets and food industries (Peters, G. M. et al., 2010). Food waste includes fruit and vegetable peelings, leftovers from meals, perishable products, etc. (Peters, G.M. et al., 2010).
- **Agricultural waste:** This comes from farming activities, such as crop residues, animal excrement, straw, dead leaves, etc. This waste can be composted, methanised or used as a natural fertiliser. This waste can be composted, methanised or used as natural fertiliser (Chan, M. T. et al., 2019; P200).
- **Green waste:** This is garden waste such as dead leaves, grass cuttings, tree branches, etc. (Wu, 2018; pp. 262). It can be composted to produce a nutrient-rich soil.
- **Industrial waste:** according to Buswell (2008; pp.117) this waste comes from industrial activities, such as food production, paper production, textile production, etc. This waste can be composted, methanised or used in biogas production. This waste can be composted, methanised or used to produce biogas.
- **Sewage sludge:** This is the solid residue from wastewater treatment plants (Liu, R. et al., 2019; p. 98). This sludge can be composted or used to produce biogas.

These categories of organic materials represent a significant proportion of the waste produced in various sectors. Their proper management is essential to reduce their environmental impact and maximise their recovery.

1.2.2. Biogas production

Biogas production is a process involving the transformation of organic matter into combustible gas (Maglwa T., 2019, p. 77) in installations known as anaerobic digesters (Belanger F., 2009, p. 5).

1.2.2.1.   Biogas production technologies

In general, biogas production technology includes the principle of anaerobic digestion and the digester to be used.

> *Basic principle of anaerobic digestion*

Anaerobic digestion is a bacterial degradation of organic substrates (wastewater, agricultural waste and livestock effluent) that occurs spontaneously in oxygen-poor natural environments (marshes, bovine digestive tracts, wastewater pits) (www.water.gov.ma). Because of its gaseous production and biological origin, anaerobic digestion is frequently referred to as 'bio-methanisation' (Aleke, 2016, p. 71). This process takes place in a reactor that produces biogas made up of 50 to 70% methane (CH_4), 30 to 50% carbon dioxide (CO_2) and other trace gases (water vapour, H2S) in addition to the effluent output, known as digestate (Dupont Jeannine, 2020.).

> **Biogas digesters**

A digester or biodigester is an anaerobic system that converts organic matter into biogas and digestate (Yan at al., 2022, p. 32). It is a promising technology for producing biogas from organic waste (Marchaim at al., 1994). Arnaud C., (2021, p. 7), defines a biodigester as a closed reactor in which micro-organisms degrade organic matter in the absence of oxygen, thereby producing biogas.

Operation of the biodigester :

According to Portillo (2023; p5), the biodigester works thanks to the action of anaerobic micro-organisms, mainly methanogenic bacteria, which break down the organic matter in different stages of degradation. This occurs in an oxygen-free environment created inside the biodigester, according to Gracia (sd; p. 11). According to G. Sajeesh et al (2017, p. 95), bacteria first degrade soluble matter, then specific matter in a multi-stage process to produce biogas.

Types of biodigester :

There are several types of biodigester available, each with its own advantages and disadvantages. Here are some common examples:

1. **Continuous flow biodigesters**: These biodigesters are fed continuously with organic waste. A. Pandey et al (2019), indicate that they offer constant biogas production, making them suitable for large-scale commercial operation.

2. **Batch-flow biodigesters**: Unlike continuous-flow biodigesters, these systems are fed periodically with organic waste. S. R. Khanal (2019), specifies that they are more suitable for small farms or households because they need to be filled and emptied periodically.

3. **Fluidised bed biodigesters**: These biodigesters use a suspended bed to ensure efficient mixing of organic waste and micro-organisms. R. Mohee et al (2014) report that they are known for their superior performance in terms of biogas production and reduction of hydraulic retention times.

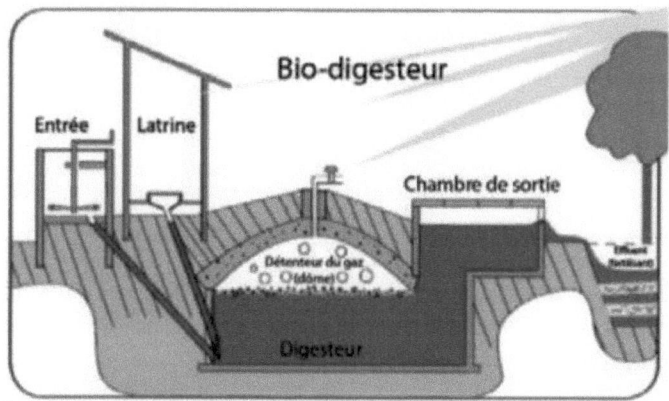

Figure 1: A family biodigester in the ground.

Source: Sama, H. & Tiabri Thiombiano, S., 2018.

It is important to note that the selection of the type of biodigester will depend on local factors such as the type of waste available, the size of the farm, water resources, etc. The costs of construction, operation and maintenance should also be taken into account, as well as the technical skills available. The costs of construction, operation and maintenance should also be taken into account, as well as the technical skills available.

Advantages of the biodigester

The biodigester has many advantages, both from an environmental and socio-economic point of view. Here are some of the most important advantages of the biodigester:

- **Efficient conversion of organic matter into biogas**

The biodigester is an efficient way of converting organic waste into biogas. According to Koenraad Vanhoutte (2016, PP841), author of the study "Biodigesteur domestique: un outil pour une gestion durable de la biomasse organique", the rate of conversion of organic matter into biogas can reach up to 70-80% in a biodigester.

- **Reducing organic waste and greenhouse gas emissions :**

By using a biodigester, organic waste is treated efficiently, preventing it from accumulating in landfills and decomposing, thereby emitting harmful greenhouse gases (PNB-BF, 2019, P4). According to a study by Suzanne L. et al, (2017; PP389), the use of biodigesters significantly reduces greenhouse gas emissions from the anaerobic decomposition of organic waste.

- **Biogas production, a renewable energy source**

The biogas produced by the biodigester can be used as a renewable energy source. Shyam S. et al, 2007; PP1211), said that biogas is composed mainly of methane, a combustible gas that can be used to produce electricity, heat or fuel for vehicles.

- **Generation of digestate, a nutrient-rich natural fertiliser**

The biodigester produces. Digestate, as pointed out by Teodorita (2008, p. 7), contains a high concentration of nutrients, such as nitrogen, phosphorus and potassium, making it a high-quality natural fertiliser. Schwope (2016; p. 13) points out that the use of digestate in agriculture promotes soil fertility and reduces dependence on chemical fertilisers.

Finally, the biodigester offers a number of advantages, including the efficient conversion of organic matter into biogas, the reduction of organic waste and greenhouse gas emissions, the production of renewable energy and the generation of digestate, a nutrient-rich fertiliser. These advantages make it an attractive solution for the sustainable management of organic waste and the production of clean energy.

Factors influencing biodigester performance

- **Composition of the waste fed to the biodigester**

The performance of a biodigester can be influenced by a number of factors, including waste composition, temperature, pH, mixing and agitation methods, and hydraulic load control (Laperriere, 2017; p. 19).

- **Optimum temperature and pH for anaerobic biodigestion**

Temperature and pH are also key factors to consider. According to research by Silva et al (2019, p. 8), anaerobic biodigestion generally works best at temperatures between 35°C and 55°C. In addition, the pH must be maintained within a specific range (generally between 6.5 and 8) to favour the activity of the methanogenic bacteria responsible for biogas production (Malik et al., 2020; p. 11).

- **Mixing and agitation methods to promote the biodigestion process.**

With regard to mixing and agitation methods, several studies, such as that by Kumar et al. (2019, p. 93), have shown that periodic agitation and the use of efficient mixing systems improve biodigestion by promoting better distribution of microorganisms and nutrients in the biodigester.

- **Hydraulic load control for optimum performance.**

Controlling the hydraulic load is also essential for optimising biodigester performance. According to a study by Ahring (2003, p. 2) in the journal Water Research, excessive hydraulic loading can lead to fat accumulation and inhibition of microbial activity, while insufficient hydraulic loading can reduce biodigestion efficiency.

It should be noted that these factors can vary depending on the type of biodigester used, the scale of production and the specific environmental conditions. It is therefore essential to take these factors into account and adjust them according to the specific needs of each biodigester.

Constraints and challenges of the biodigester

Biodigesters can face certain constraints and challenges that can affect their performance and long-term use (ONAS, 2013; p. 42). Here are some of these factors, with author references for each point:

1. **Availability and quality of organic waste**: The composition of the organic waste used as feed in a biodigester can play a crucial role in biogas production (RECORD, 2003; p. 36). Studies have shown that the balance between carbon/nitrogen ratio, moisture content, acidity and organic load can influence biodigester productivity (Borowski et al., 2019). It is therefore important to select organic waste sources carefully to maximise biogas production.

2. **Capital and maintenance costs**: the installation and maintenance of a biodigester can involve significant initial costs (Rodrigues and Preston, 1996; p. 9), including the construction of the structure, the purchase of equipment and the installation of biogas capture and storage systems. Studies have highlighted that the cost of construction and maintenance can vary depending on the size of the biodigester and the technologies used (Nges et al., 2015). A thorough assessment of potential costs is therefore required before implementing a biodigester.

3. **Risk management**: biodigesters produce biogas, which is flammable and potentially dangerous if not properly collected and stored. Strict safety measures must be taken to minimise the risk of gas build-up and explosions. Studies have highlighted the importance of adequate biodigester ventilation and the use of safety systems for biogas control (Al Seadi et al., 2008).

4. **Technical skills requirements**: the operation and maintenance of a biodigester requires specific technical skills to ensure that it runs smoothly. This may include knowledge of anaerobic biodigestion processes, waste management, monitoring of key parameters such as temperature and pH, and the ability to troubleshoot any technical problems. Training and capacity building programmes may be required for biodigester operators (Zhao et al., 2018).

Finally, the availability of organic waste, investment and maintenance costs, risk management and technical skills are critical aspects to be taken into account when implementing and operating a biodigester. The scientific literature offers solutions and recommendations for mitigating these constraints and overcoming the challenges associated with the biodigester.

1.2.2.2.   Types of substrate for biogas production

In biogas production, substrate refers to the organic matter used as raw material for anaerobic fermentation (Laperriere W. at all., 2020, P3). This is the main source of energy for biogas production.

Several types of substrate can be used provided that they are biologically degradable in anaerobic factors within a short timeframe (RECORD, 2022, P25).

To be methanised, a substrate must be rich in biodegradable organic matter, excluding woody matter (HO B., 2021, P66) and not contain any digestion-disrupting elements (undesirable, inhibitors, etc.) (RECORD, 2009, P11).

> **Animal faeces are a** source of methanogenic bacteria and have the power to cause disease.
but have limited methanogenic potential (F. T. DIOP at all. 2015, P2).

Figure 2. Cattle dung at the Kituku slaughterhouse in Goma

Photo taken on 8 May 2023 at 09:21.

> **Agricultural plant matter**: Produced on the farm (silage from intermediate or forage crops, residues) or silo waste, plant matter has interesting methanogenic potential and can easily balance a ration.
> **Municipal and catering waste**: grass cuttings, leaves (watch out for woody plants), sewage sludge, catering waste.
> Agri-food industry waste: from industries that process milk, meat, fruit, vegetables and breweries. This waste generally has good methanogenic potential, but requires appropriate storage (AILE, 2022).

Each type of substrate has advantages and disadvantages in terms of quality and profitability. The choice of substrate often depends on local availability and the economics of production.

Table I. M.0 resources available in the city of Goma

N°	RESOURCES AVAILABLE FOR THE BIOGAS PRODUCTION IN THE TOWN OF GOMA	RESOURCE LOCATION	USE OF THE RESOURCE IN BIOGAS PRODUCTION
01.	Organic waste	- Households, - Markets, - Restaurants - and other sources	The city produces a considerable amount of organic waste. This waste can be used as a raw material for the production of biogas.
02.	Animal waste	- Slaughterhouses - Backyard livestock	Goma is home to a number of livestock farms and slaughterhouses, which generate large quantities of animal waste. This waste can be collected and used to produce biogas.
03.	Industrial effluents	- Food agri-food	Goma also has industries that generate organic effluents, such as food waste or production residues. These effluents can be used to produce biogas.

Source: Our investigations; August 2023

1.2.3. Biogas

Biogas can be produced from all types of organic waste (Beth D., 2008, P1), but also from unsold supermarket produce and catering waste (Bourdin S., 2020, PP61).

1.2.3.1. Possible uses for biogas

Biogas is a versatile source of energy that can be used in a variety of ways. Here are two common uses for biogas:

> *Domestic use by households*

The most common use for biogas is in cooking (Beline F., 2007, P7). For Batside G. (2014, P3), biogas can be used as a fuel in gas cookers, replacing fossil fuels such as natural gas or propane. This reduces dependence on non-renewable energy sources and cuts greenhouse gas emissions (Lavasseur P., 2023, P20).

> *Use after transformation into electric current*

According to Quinton E. (2008, P11), biogas can be used to produce electricity using biogas generators or adapted internal combustion engines. This process is known as cogeneration (Welter C., 2010, P2), because it produces both electrical energy and heat. The electricity produced can be used to power industrial facilities, residential buildings, farms or even fed into the electricity grid.

> *Use for heating*

Biogas can also be used to produce heat directly. This heat can be used to heat buildings, dry crops, heat greenhouses, treat water, etc (Chareyron D. at al., 2021, P5). It can be used in specially designed boilers to burn the biogas and produce heat.

> *Use as a natural augaz additive*

Another use for biogas is direct injection into the existing natural gas network (HotRot, quoted by Amarante J., 2010, p. 19). After a purification process to remove impurities, biogas can be mixed directly with natural gas and used in the same applications as the latter, such as heating, cooking, power generation, etc. (CRAAO, 2008, p. 15). This use of biogas promotes the transition to a greener, more sustainable economy by using a renewable energy source (atee Club, 2014, p. 3).

It should be noted that the uses of biogas may vary depending on the resources available and local needs. Some facilities may also use biogas to produce alternative fuels, such as biodiesel or compressed natural gas (CNG) for vehicles. These uses depend on the infrastructure and regulations in place in each region.

Partial conclusion

To conclude this literature review, we have explored in detail the essential concepts associated with the production of biogas from organic matter. We have accurately defined organic matter and highlighted its importance in the context of biogas production. We have also examined the theoretical approach underlying this process, analysing the scientific principles and techniques used to optimise production.

Chapter II
RESEARCH METHODOLOGY

The first section of this chapter will be devoted to presenting the study environment, the city of Goma. The second section will focus on the materials and tools we used to carry out our research. We will describe the equipment, the measuring instruments, the sample collection materials and any other elements essential to our experiment, and we will discuss the experiment itself. We will explain the different stages we followed to collect the samples, carry out the measurements and conduct our research into biogas production.

Finally, we present the general methodology we have adopted for this study.

11.1. PRESENTATION OF THE STUDY ENVIRONMENT

11.1.1. Geographical location

The town of Goma is located south of the Equator (BYANIKIRO, 2010, P27) between 1° 40' 26.7 "and 29° 13' 42.4" longitude East (Musorongi, 2014, P32). It is bordered to the north by Nyiragongo territory, to the south by South Kivu province, to the west by Masisi territory and to the east by the Republic of Rwanda (INS, 2015, P4). It covers an area of 66,824 km^2, or 11% of the province of North Kivu (Safari, 2014, P23). The town, built at the foot of the NYIRAGONGO and KARISIMBI volcanoes, is entirely covered in volcanic soil with a relatively flat topography. "Its altitude varies between 140m at the edge of Lake Kivu and 2,000m at the point where it joins the BUKUMU community" (Kamalebo, 2013, P13). The town has a single highest point, Mont Goma (CGES, 2017).

11.1.2. Administrative and demographic aspects

11.1.2.1. Administrative aspects

Once the capital of the North Kivu sub-region, Goma is now the capital of North Kivu province.

Subdivided into two communes, it is a decentralised entity headed by the mayor, assisted by a deputy mayor. The commune of Goma has seven districts: Les Volcans, Mikeno, Katindo, Kyeshero, Mapendo, Himbi and Lac-Vert. The commune of Karisimbi lies to the north of Goma and to the south of Nyirangongo territory. It is also subdivided into eleven districts: Kahembe, Majengo, Mabanga Nord, Mabanga Sud, Murara, Kasika, Katoyi, Ndosho, Mugunga, Bujovu and Virunga. Each commune is headed by a Bourgmestre (mayor) assisted by a deputy. Each commune is subdivided into different neighbourhoods, which are in turn subdivided into Cellules, and the latter are also subdivided into Avenues (Kanambe JUAKALY, 2012, P19).

Moreover, since its creation, the city of Goma has been led by different personalities who have succeeded each other in the following way (WITANDAY, 2011, P21 and CHEKA, 2021):

1. Kanga GUZANGAMANA from 1989-1991,
2. Migale Mwene Malibu in 1991 for just a few months
3. Athanase Kahanya Kimuha Tassi from 1991 to 1993
4. Mashako Mamba Sebi from 1996 to 1998
5. Kisuba Shebaeni from 1996 to 1998
6. Fr Xavier Nzabara Matsesa from 1998 to 2005
7. Polydore Wundi Kwavwira from 2005 to 2008
8. Rachid Tumbula from 2008 to 2010
9. Jean Busanga Malihaseme from 2011 to July 2012

10. KUBUYA NDOOLE Naason from 14 July 2012 to August 2015
11. Dieudonne MALERE MA-MICHO from August 2015 to 2016
12. Timothee Mbarushimana from 2016 to 2019
13. Celestin Nsabayesu from 2019 to 2021
14. Francois Kabeya from 2021 to present

Town Hall organisation chart

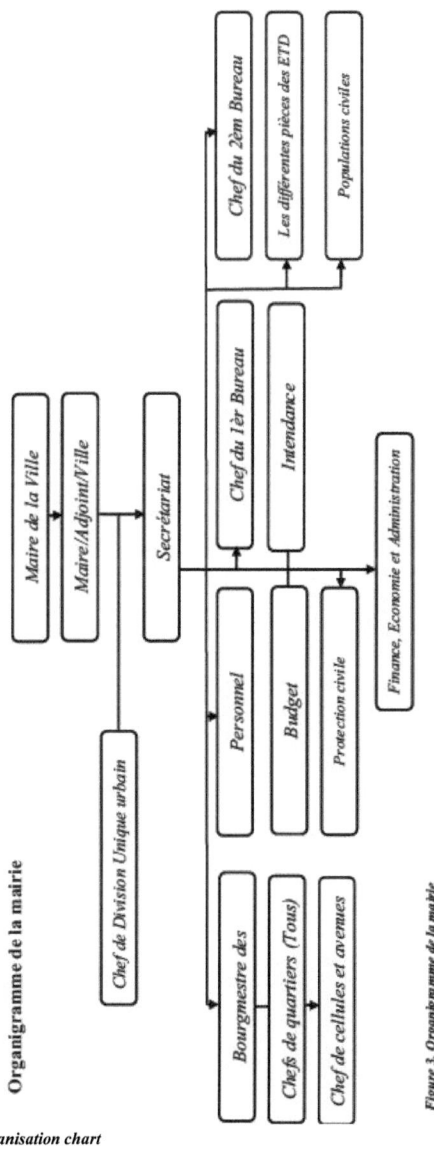

Figure 3: Town Hall organisation chart

Source: M. M'PINDA, 2010, P11

II.1.2.2. Demographic aspect

The city of Goma has a high demographic concentration characterised by ethnic heterogeneity (Rachid, 2014, P39), coming from territories ravaged by tribal-ethnic wars over the past decade and from other provinces of the DRC (UN-Habitat, 2017, 23). In addition to this reason, according to the INS (2018, 34), there is also an economic one, due to the fact that the geographical position of the city of Goma attracts the attention of immigrants who engage in trade. There has therefore been rapid population growth in the town of Goma (Muhima, 2020, P31). Mincho (2021, P12) believes that the geoclimatic factors that characterise the city of Goma cannot be denied.

In 1998, the urban population of the city of Goma was 318,173 and after just 12 years it has risen to 628,352 (PASA 2016). The latest statistics from Goma town hall (Mairie de Goma, 2020) give 1,133,135 as the current population total.

Table II. SUMMARY TABLE OF THE NUMBER OF CONGOLESE, FOREIGN NON-REFUGEE AND FOREIGN REFUGEE INHABITANTS IN THE CITY OF GOMA

SUBDIVISION ADMIN.	POPULATION CONGOLAISE					POPULATION ETRANGERE NON REFUGEE					POPULATION ETRANGERE REFUGIEE					POPULATION TOTALE				
	H	F	GA	FI	TOT	H	F	GA	FI	TOT	H	F	G	F	TO	H	F	GA	FI	TOTAL
COMMUNE DE GOMA																				
Q-LES VOLCANS	4268	4052	3364	3364	15048	538	108	66	13	725	-	-	-	-	-	4806	4160	3430	3377	15773
Q-MIKENO	8283	8391	9558	9441	35673	7	7	4	1	19	-	-	-	-	-	8290	8398	9562	9442	55773
Q-MAPENDO	5572	5688	10299	10454	35673	15	16	27	32	90	-	-	-	-	-	5587	5704	10326	10486	32103
Q-KATINDO	7008	6669	8270	2591	30538	15	20	9	17	61	-	3	-	9	8	20	7023	6692	8288	8616	30619
Q-HIMBI	2315	3575	12508	3430	51825	7	9	-	-	16	-	-	-	-	-	12822	13584	12508	13430	51844
Q-KYESHERO	4797	6288	25214	28057	84336	35	13	-	-	48	-	-	-	-	-	14832	16301	21465	28057	84704
Q-LAC VERT	16542	18462	21907	21020	77931	14	2	-	-	16	-	-	-	-	-	16556	18464	21907	21020	77947
TOTAL GOMA	87855	125	91120	94357	27387	631	175	106	63	975	-	3	-	9	8	20	69416	73503	91535	94428	348682
COMMUNE DE																				
Q-MURARA	6975	7756	10126	1202	36059	3	2	4	1	6	-	-	-	-	-	6978	7758	10130	11203	36069
Q-KAHEMBE	4363	5311	8871	9275	28820	3	2	3	3	11	-	-	-	-	-	4366	5313	8874	9278	25851
Q-MAJENGO	2152	3760	16833	18446	61191	2	1	2	4	9	-	-	-	-	-	12154	13761	16835	18450	61200
Q-VIRUNGA	3509	3687	4763	4882	16841	-	-	-	-	-	-	-	-	-	-	3509	3687	4763	4882	16841
Q-MABANGA NORD	7291	9045	10162	3753	40253	7	5	1	-	13	-	-	-	-	-	7298	9050	10163	3755	40266
Q-MABANGA SUD	19029	4140	7653	30464	61286	5	2	-	1	8	-	-	-	-	-	19034	24142	7653	30465	61294
Q-KASIKA	10505	11372	13943	14703	50523	-	1	-	2	4	-	-	-	-	-	10505	11373	13945	14704	50527
Q-NDOSHO	10519	1693	21497	2449	68266	-	-	-	-	-	-	-	-	-	-	10519	11693	21497	24497	68206
Q-MUGUNGA	2703	3419	5510	6058	17690	-	-	-	-	-	-	-	-	-	-	2703	3419	5510	6058	17670
Q-BUJOVU	5400	6190	10468	11122	33180	1	2	-	-	3	-	-	-	-	-	5491	6192	10468	11122	33183
Q-KATOYI	16568	9165	22031	24498	82762	13	13	6	1	33	-	-	-	-	-	16581	19178	22037	24499	82795
TOTAL KARISIMBI	9014	5358	22985	49403	38811	34	28	11	91	-	-	-	-	-	-	99048	115366	129875	149413	63714?
TOTAL VILLE DE						665	203	124	74	1066	-	3	-	9	8					133 135

Source: **MAIRIE DE GOMA (2020, P20), population census report for the city of Goma.**

11.1.2.3. Energy aspect

In the Democratic Republic of Congo, the use of wood fuel, and charcoal in particular, is in the majority in urban areas (BIKA, 2015, P9). And according to the CNE (2011, P14), more than 99% of households in Goma use charcoal regularly to cook their food. The average daily consumption of wood fuel by a resident of Goma is 0.43 kg of charcoal and 0.02 kg of wood, i.e. 3.50 kg of wood equivalent (CIRAD, 2020; P:1).

The population of the city of Goma uses several energy sources (INDP, 2022, P35), and according to the Africa Energy Outlook (2019, P202), people attach more importance to those that meet the most recurrent needs, such as cooking. Here are some of the energy sources used in Goma:

1. Charcoal

According to CIRAD (2020, P5), charcoal is the most widely used energy source in the city of Goma. In 2015, an estimated 65% of the population used this source of energy, mainly for domestic purposes or as part of informal production (WASIWASI quoted by Nixon, 2015). In the aftermath of the volcanic eruption, charcoal consumption increased to 80% of users (Nixon, 2015).

Goma's annual consumption of wood fuel amounts to 1.26 million tonnes of wood equivalent for a population estimated at 1 million, broken down into 186,000 tonnes of charcoal and 32,000 tonnes of firewood (CIRAD, 2021; P:7).

2. Lamp oil

This energy is used in particular for lighting in households in the evening and to some extent for cooking in Goma. Consumption is particularly high because a large number of households are not connected to the electricity network. Fuel stations have been set up at strategic points in the city of Goma to supply vehicles, motorbikes, etc. (DAYANA A, 2004).

3. Electrical energy

Most of the electricity consumed in Goma comes from the Ruzizi hydroelectric power station in South Kivu province and the Virunga plant at Matebe in Rutshuru territory. Another source of electricity is the Nuru photovoltaic farm, which supplies the Ndosho district and parts of Mugunga, Lac Vert and Nyiragongo territory.

Some households use solar panels to be self-sufficient in energy, while others use two or more energy sources to compensate for the mismatch.

11.1.2.4. Economic aspect

Unlike other regions whose economy is based on mining resources, the province of North Kivu in general and the city of Goma in particular rely more on agricultural and customs resources. A large proportion of the town's inhabitants are involved in marketing agricultural produce; most economic transactions are conducted in foreign currency, due to the instability of the local currency.

J Trade

There is a wide variety of trade, ranging from small-scale trading in front of plots of land to medium-scale trading in markets and shops to large-scale import and export trading. The products sold come from a variety of sources, depending on the ability of the traders involved. The closeness of the towns of Goma and Gisenyi has led to the development of a considerable trade movement, especially for basic necessities such as maize, sorghum, potatoes, tomatoes, etc. Shops, mini-foodstores and pharmacies make it easier to supply the people of Goma with all kinds of products. Shops in the town centre import, export and distribute (sell) products from neighbouring countries and elsewhere. Several small and large markets are hidden here

and there in the city of Goma. Some are regulated administratively, while others known as KASOKO are sometimes held without the approval of the local authorities.
In the town of Goma, there are eight officially recognised markets, apart from the clandestine ones.

Table III. Walks in the town of Goma

Walk	Date of creation	Municipality
Virunga	1972	Karisimbi
Mikeno	1973	Goma
Kahembe	1995	Karisimbi
Kituku	1995	Goma
Olive lembe (Alanine)	1996	Goma
Katoyi	1997	Karisimbi
Majengo	1997	Karisimbi
Nyabushongo	-	Karisimbi

Source: Goma town hall, 2020

These public contracts are managed respectively by the town hall and the communes.
Town hall: Virunga and Mikeno walks
Commune of Karisimbi: Marche Kahembe, Katoyi and Majengo
Goma commune: Marche olive lembe, Kituku and Nyabushongo.

J Transport

We can attempt to review the various modes of transport, highlighting in passing their recent evolution and development, the main aim of which is to maintain and increase the income of transport operators, but which often results in a reduction in costs that ultimately benefits the user. Several transport companies move people and goods between Goma and the surrounding area. Transport in the city of Goma covers air, lake and land transport.

Air transport via Goma international airport enables people and goods to be moved both nationally and internationally. With the reunification of the country, just after the so-called rebellion period, a number of private companies that operated the Goma air route either ceased to operate or no longer serve the city. These include companies such as VIRUNGA AIR, CHARTER, SUNAIR, AIR BOYOMA, Goma AIR, CAGL, AIR KIVU and others. Among the companies that continue to make their presence felt in the air transport sector are some old names such as TMK, to which others have been added such as WIMBI DIRA, AIR WAYS, HEWA BORA, AIR LINES, MANGOMAT, etc (Yusuf MUHINDO, 2020).

Lake transport is carried out on Lake Kivu, facilitating travel between the towns of Goma and Bukavu in the province of South Kivu. It is a fairly important route because of the number of people and the quantity of goods exchanged between the inhabitants of these two towns. The main companies in charge of operating this waterway are: the Societe Nationale des Chemins de Fer du Congo (SNCC), and private companies such as: RAFIKI, SALAMA, TMK, IHUSI, MV IKO, KARISIMBI, EMMANUEL, MUGOTE, etc.

11.1.2.5. Environmental and climatic aspects

The city of Goma is confronted with environmental problems ranging from the low number of paved roads, natural hazards (volcanic eruption), poor solid waste management and air pollution, to inadequate management of wastewater and sewage sludge, the deterioration of urban roads and the shoreline of Lake Kivu, etc (Provincial Planning Ministry, 2017, P20). All these problems have a definite impact on the living environment and health of the city's population, and even on the biodiversity of Lake Kivu.

The sanitation report (management of domestic waste, rubbish and drainage water) reveals many shortcomings that are of concern to the municipal authorities, such as the slowness of waste removal (Mercycorps, 2022, P3), from pre-collection sites to undeveloped final landfill sites (CIRHAKARHULA, 2020, P10). It is worth noting the many environmental nuisances and damage associated with waste (WITANDAY, 2009, P14). The presence of methane gas in Lake Kivu poses a permanent risk of pollution and poisoning for the city's inhabitants (CGES, 2017).

The town of Goma lies entirely in the inter-tropical zone (Lebrun, 1959, P42). The climate is generally mild thanks to the winds blowing off Lake Kivu and its altitude (1670m). There are two seasons (Bagula, 2011, P17):
- The dry season: the long and short mountain dry season;
- The rainy season: the long and short mountain rainy season.

It also has a temperate climate due to its altitude and the presence of Lake Kivu (Mairie de Goma, 2020, P13).

II.1.2.6. Socio-cultural aspect

The city of Goma is a crossroads where the indigenous population is growing rapidly (BUCHEKABIRI, 2013, P15), but it is also mixed with foreigners, notably Lebanese, Belgians, Indians, Ugandans, Burundians, Rwandans, etc (Rugurike, 2022, P19). Goma's diverse population has a wide range of socio-cultural behaviours.

11.2. MATERIALS AND TOOLS USED, AND EXPERIMENTATION
11.2.1. Tools and equipment used
11.2.1.1. Search tools

The data collection process for our research is based on two main elements: the first is documentary analysis, the second is fieldwork. Overall, the following tools were used:
- ❖ Written documents
- ❖ The survey questionnaire ;
- ❖ The logbook ;
- ❖ Cards, certificates and pens ;
- ❖ Rolling stock to make moving around easier;
- ❖ Computer ;
- ❖ Cobo collect SPSS
- ❖ Telephone; etc.

11.2.1.2. Equipment used

We used a number of different materials to demonstrate our experiment. Here are the main ones:
- ❖ Tank
- ❖ Plastic bottles
- ❖ Flexible hose   1/8
- ❖ Valve   1/8
- ❖ Sleeve   1/8
- ❖ Nipple   1/8
- ❖ Connection
- ❖ Reducer
- ❖   Elbow%.
- ❖   T%

11.2.2. EXPERIMENT
11.2.2.1. For Methanisation
a) Principle

Methanisation is the result of complex microbial activity carried out under anaerobic conditions (Beline, 2007; P33). Anaerobic digestion is a biological degradation process that transforms complex organic substrates into molecules containing a single carbon, such as methane (CH_4) and carbon dioxide (CO_2) (Digan, 2019; P19).

The process takes place in four distinct phases, each carried out by a specialised class of micro-organisms, which develop in the absence of oxygen:

- **Stages 1 and 2: Hydrolysis and acidogenesis**

Degradation of polymers into monomers and then volatile fatty acids. This is a limiting stage that can be improved by pre-treatment, but blockages are fairly rare in practice (Afsset, 2008; P30).

- **Stage 3: Acetogenesis**

Transformation of VFAs, hydrogen and carbon dioxide into acetic acid. This stage is very rapid compared with the others (Batstone et al., Cited by Wassila, 2017; P24).

- **Stage 4: Methanogenesis**

Methane is formed either by degradation of acetic acid (70% of production) or by reduction of CO_2 by hydrogen (30% of production) (Bernet, 2020; P3). This is a limiting stage because it is highly sensitive to variations in environmental conditions (pH, temperature, toxic agents, variations in effluent concentration, etc.) (Record, 2020; p. 13). In principle, it is this phase that is controlled.

The micro-biological mechanism of anaerobic digestion is now fully understood.

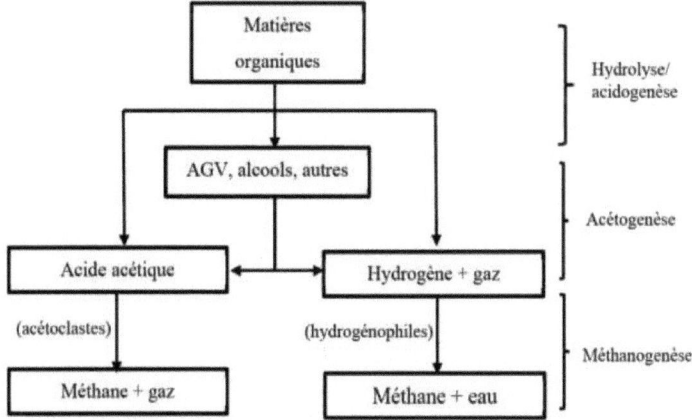

Fig. 2: Summary diagram of the main stages in anaerobic digestion (Cresson et al., 2006).

b) Theoretical factors influencing digestion

The theoretical factors involved in digestion, resulting from consideration of the biomass in relation to the substrate to be digested, are as follows: load, pH, the "temperature - residence time" pair and toxic substances (Doublet at al., 2004; p. 18).

- The load

The load of a digester is obtained by dividing the weight of volatile matter introduced by the volume of the digester (Rivard, 2015; P34). It is expressed in kg of MV/m^3 and is used to assess the quantity of nutrients introduced in relation to the biomass present, which at the end of digestion contains all the bacterial strains needed to digest the fresh sludge (BOUYAHIA, 2020; P101). We manually estimate that the digester feed should constantly respect the following proportion: 20 times more digested sludge than fresh sludge (expressed in kg of MV). In addition, any significant variation in load disrupts the digestion process and can lead to it stalling.

- **The pH**

This is one of the most important factors in the adaptation of bacterial populations (Goux, 2015; P15). Acetogenic bacteria remain active up to a pH of 5, but the bacteria responsible for methanisation are inhibited as soon as the pH drops below *6.2* (Quinton at al., 2008; P2). This is why the theoretical optimum pH range to be respected is between 6.8 and 7.2 (Amarante, 2010; P10). In the event of a drop in pH due to an increase in the concentration of fatty acids, the pH can be maintained by adding soda ash or lime to the thickened fresh sludge feed or directly to the digester (Moisse, 2020; P23).

- **The temperature**

Anaerobic digestion takes three different forms depending on temperature
- When the temperature is below 15°C, digestion is psychrophilic (Mabala, sd; p. 32),
- When the temperature is between 28 and 40°C, digestion is of the mesophilic type (Batside, 2014; p. 6) and
- When the temperature is above 45°C, digestion is thermophilic (Hess, 2007; p. 27).

- **Temperature - length of stay**

The residence time (or hydraulic retention time TRH) is the theoretical time during which the volume of fresh sludge remains in the digester (Mohamed at al., 2015; P19). It is obtained by dividing the sludge flow rate entering the digester by the digester volume (Faixo, 2020; P9).

The temperature must be kept constant to avoid disrupting digestion (Pouech, 2017; P6). Methanogenic bacteria are particularly sensitive to any variation in temperature, even of the order of 1°C per day (Marchaim at al., 1994; p. 97).

Temperature and residence time are two related factors (Rabat, 2012; p. 7). Indeed, an increase in temperature leads to the activation of acetogenesis, methanisation and bacterial growth reactions (Cuny, 2018; P8). This results in a reduction in the residence time required for stabilisation and an increase in gas production (Rabah at al., 2020; p. 20).

- **Toxic substances**

There are several types of digestion inhibitors:
o Certain cations and sulphides (Doublet at al., 2004; p. 8).
o Trace elements (Bougrier, 2005; p. 36).

It has been shown that trace elements can activate or inhibit the anaerobic digestion process (Degremont, *Memento technique de l'eau*):
o They are necessary for the growth of certain bacteria, methanogenic bacteria (Agence de l'eau Rhone Mediterranee Corse, 2012; p. 16).
o They inhibit the toxicity produced by sulphides (Frata, 2016; p. 2).
o They allow the formation of phosphates and the aggregation of bacteria (Comte quoted by Jaziri, 2012; p. 11).

However, in too high a concentration, trace elements inhibit the growth of bacteria and therefore the production of :

- Methane (INERIS, 2006; p. 12);
- Certain organic compounds: cyanides, phenols, phthalates (Soa, 2016; p. 26);
- Antibiotics (Khadra et al., cited by El Fels, 2014; p. 2);
- Detergents. (Frata, 2016; p. 2); etc.

c) Urban wastewater treatment plants

The anaerobic digestion process (sludge treatment and stabilisation process) is applied to concentrated sludge after primary or secondary wastewater treatment (Zhang, 2011; p. 8).

d) Methanisation of industrial effluent

Methanisation of industrial waste or effluent is competitive when the biodegradable matter, expressed as COD (chemical oxygen demand), is sufficiently high (Bautitsa, 2019; p. 62). It is therefore used on various substrates, effluents or sludges from: distilleries, breweries, fruit and vegetable canneries, dairies, slaughterhouses, sugar refineries, paper mills, pharmaceutical industries, starch, citric acid and malt production plants, textiles, etc. (Rabah at al., 2020; p. 23).

In most cases, biogas is used directly for thermal purposes on the production site (boilers, turbines, engines). Cogeneration is relatively underdeveloped.

e) Methanisation of municipal biowaste

Biowaste is green waste, kitchen waste and paper and cardboard from local authorities. **The biogas is mainly recovered by cogeneration using gas engines (Marois, 2014; p. 81).** The electricity is sold to the grid, and the heat is either used for the process or exported (Apolit at al., 2021; p. 21).

f) Methanisation of livestock effluent

Livestock effluent covers all animal production: pigs, cattle, poultry.... (Steinfeld, 2009; p. 110).

Farmyard manure is mainly pig and cattle slurry and manure (Cuny, 2015; p. 6).

The issue is different when talking about individual units (on-farm biogas) and centralised units (Suraud, 2018; p. 5). The context of on-farm biogas is fundamentally different and concerns small installations of no more than a hundred kWe (Ernest and Young, 2010; p. 59). The biogas is converted into heat and electricity using **cogeneration on a gas engine or dual-fuel engine** (Pinard, 2011; p. 42).

g) Co-digestion

All the substrates listed above can be treated as a mixture by methanisation (Bel at al., 2016; p. 15). This is known as co-digestion. These units are new applications that are still under-represented.

h) Waste storage facility (ISD) : landfill methanisation

The waste is compacted and buried in an area that is impervious to any external intrusion (water, air) (Couturier at al., 2001; p. 17). The methanisation process therefore occurs 'spontaneously' inside waste storage facilities (Toine at al., 2014; p. 64).

Thermal recovery exists in significant proportions (on-site production: steam, drying, dehydration of leachates) (Garcia-Bernet at al., 2017; p. 13) but is limited by the absence of local outlets (external recovery: delivery of heat to an urban network or an industrial plant) (Perrault, 2007; p. 4). Cogeneration (production of electricity and heat) is mainly used for heat recovery (Bastide, 2014; p. 2).

II.2.3. Gasification

a) Principle

Gasification is a process that converts carbon material into simple fuel gases (AFP, 2011; p.

7).
Gasification takes place at low and medium temperatures with little oxygen (reduction) (Carlesi, 2012; p. 129). In conventional combustion processes for solid carbon materials, the various phases of drying, dissociation of volatile compounds and carbon, and total oxidation of gases, liquors and carbon take place in the same reactor (Teixeira, 2012; p. 9).
One of the advantages of gasification is that it separates these phases, with the gases being burnt in specific equipment (Georgio, 2012; p. 13).
The general process of gasification can be summarised as follows:
- The first stage is **drying, at around 100-200°C,** with evaporation of the water (EL FELS, 2014; p. 184).
- The second stage is **pyrolysis, which is** a thermal decomposition of the substrate in the absence of oxygen, and produces gaseous and liquid hydrocarbons (simple and complex) and solid carbons (Kohler, 2009; P12). Overall, the reactions are as follows:

$C_xH_yO_z \rightarrow C + CO + CO_2 + H_2O +$ **Hydrocarbons. Source:** De Steene, 2014; P53

Between 200 and 300°C, a small quantity of oxygen gases (CO_2, CO) and hydrocarbons is produced.
Between 300 and 400°C, hydrocarbon production increases sharply, both in gaseous form and in condensable form (light tars, methanol, etc.) (Mhemed, 2021; p. 42). Hydrogen also appears at this stage (Huchon, 2021; p. 50).
Between 400 and 500°C, the hydrocarbon phase also generates large quantities of gaseous hydrocarbons and thick tars (Nozahic, 2008; p. 44).
Between 500 and 700°C, gas production decreases (Record, 2009; p. 20). The gas is mainly composed of hydrogen and hydrocarbons, with significant quantities of carbon oxides (Elhachemia, 2022; p. 102).
Between 700 and 1,000°C, the solid residue is composed mainly of carbon, with most of the volatile fraction having been released (Tanoh, 2021; p. 48). The production of gas, composed mainly of hydrogen with few carbon oxides, is low, as is the production of tar, composed mainly of polynuclear aromatic hydrocarbons (Rabah and Yasesr, 2020; p. 26).
During these different phases, numerous reducing reactions take place between the coal, hydrogen, carbon dioxide and water vapour, to produce carbon monoxide and methane (Zhang, 2011; p. 29).
At around 1,000 to 2,000°C, oxidation reactions take place in the presence of oxygen, steam (gas-water reaction) and hydrogen (methanisation): this is the so-called **gasification** phase. Carbon is converted into carbon oxides (CO and CO_2) (Garcia-Bernet at al., 2017; p. 3).
Three main types of gasifier have been identified (Iwunze, 2021; p. 27):
- **a fixed bed: co-current or counter-current ;**
- **he fluidised bed: in suspension or circulating ;**
- **mixed double storey.**

In conventional fixed-bed or counter-current gasification furnaces, there are different zones, from bottom to top:
A zone where air and water vapour are introduced into the ash layer, forming a bed where the temperature rises from 70 to 1100°C (Balland, 2016; p. 82).
- A highly exothermic oxidation zone, where coal from the upper layers reacts with oxygen and water vapour, generating a temperature rise to 1100°C and gases loaded with CO2, CO and H2 (Nozahic, 2008; p. 11).
- An endothermic reduction zone, where these gases react with carbon and hydrogen to

form CO (Ounoughene, 2015; p. 288).
- A pyrolysis zone, during which hot gases distil volatile compounds from the dry substrate (Morin, 2014; p. 97).
- A drying zone, from which the gases are evacuated along with the water vapour, and fed by the fresh substrate.

Depending on the operating conditions in terms of pressure, temperature and oxygen content, different types of thermolysis can be carried out (Antonini at al., 2004; p. 56).

b) The main factors of good conduct
- **Feeding and conditioning biomass**

The critical specifications of the products to be gasified mainly concern the size of the particles introduced into the reactor, and their humidity (Lopez, 2013; p. 44).

- **Particle size**

Reactors generally operate with particles between 10 and 100 mm in size, allowing hot gases to pass through without generating excessive pressure drops.

Depending on the nature of the biomass, it must therefore be either ground or compacted (sawdust, etc.) (Arena cited by Groleau, 2019; P23). Biomass fractionation also makes it possible to reduce the content of nitrogen and alkaline products (sodium, potassium, etc.) (Fethya, 2015; p. 33).

In principle, several processes are less sensitive to granulometric parameters, such as fluidised bed reactors and processes with separate pyrolysis stages (Procedes, 2001; p. 24).

- **Drying**

The level required depends on the type of technology:

Gasification requires a moisture content of less than 10-15%. Fresh biomass contains between 50 and 60% moisture, and when it is dried in the open air, it generally drops to around 20% when sufficient time has elapsed (Carlesi, 2012; p. 28).

Some gasification technologies allow direct feeding of wet biomass:
- 40 to 50% with counter-current reactors (Assouah at al., 2005; p. 150);
- 15 to 20% with co-current reactors (Goudeau, 2001; p. 2);
- 60% for fluidised beds with internal circulation (Batrak, 2005; p. 5).

c) Gasification applications

According to Bedina (2021; p. 43) a large number of products can be gasified, and the list below enumerates the materials that have been the subject of work:

- By-products from the wood processing industry:
- forestry chips
- waste wood
- wood waste from secondary processing industries
- sawdust
- contaminated wood
- barks
- Agricultural or agro-industrial by-products:
- rice straw
- wheat straw
- corn raids
- walnut shells
- cotton waste
- poultry droppings

- Municipal and industrial waste:
- municipal solid waste
- combustible fraction of municipal solid waste (RDF: Refuse Derived Fuel)
- sewage plant sludge
- hospital waste
- tyres uses
- waste from the upkeep of green spaces
- Various energy resources :
- heavy tars from oil refineries
- coal
- natural gas
- peat
- etc...

II.3. METHODOLOGY

Methodology is the set of rules and approaches adopted by a researcher during his or her research work to reach one or more conclusions (Robert K. Yin, 2014, p. 176).

11.3.1. Research framework

The project, entitled "Biogas production in the city of Goma: feasibility study", is part of the renewable energy and waste management sector. It involves
1 analysis of the technical, economic and environmental viability of setting up biogas production facilities using organic waste in the city of Goma. The study therefore covers aspects such as waste collection, the biogas production application, environmental benefits and risk management.

11.3.2. Type of search

The type of research chosen for this dissertation is an applied study, which is an experimental investigation undertaken to acquire new knowledge with a practical aim (Kitaganya and Sekimonyo. 2023; p. 32); the aim of which is to produce biogas in the town of Goma, by examining its feasibility and the issues involved.

11.3.3. Methods and techniques

BRIMO.A (1971; p. 22), defines a method as a set of intellectual operations by which a discipline seeks to achieve the truths it pursues, demonstrates them and verifies them; and for ROUVEYRAN quoted by YEO (2005; p. 1) a technique as "a proven, well-established, precise and reproducible operating procedure: a sort of cooking recipe, it describes in detail the operations necessary to obtain the best possible result, as well as their conditions of execution...".

11.3.3.1. Methods and techniques used to collect data

Data collection is a process that helps to gather information in order to make direct observations designed to answer open questions (Louana Lelong, 2023; p. 2). In other words, it is a systematic approach that involves gathering and measuring different information from different sources. This approach helps to obtain an overview of an area of interest.

A) Methods used to collect the data :

◆ **Empirical method:** empirical research represents research based on observation and experience. The empirical study gathers information called "empirical data". After analysis, this data must enable the researcher to test and respond to one or more initial hypotheses (Gaspard Claude, 2019; p. 15). This method enabled us to bring together the different opinions of our sample in order to gather objective data.

B) Techniques used in data collection :
- **The documentary technique**: According to GRAWITZ quoted by Vestine MUKARUKUNDO (2009; p. 46), the documentary technique consists of a systematic search of everything that is written in connection with the field of research. These include books, dissertations, reports, lecture notes, websites, etc. In our work, this technique consisted of gathering information from reliable sources such as books, scientific articles, government reports, etc. This technique was used to identify the most relevant sources of information. This technique was used to obtain basic data on the description of the city of Goma, biogas production, existing technologies, as well as similar studies that have been carried out in other regions.
- **Interview and survey techniques**: the term 'survey' covers a range of technically different forms of investigation, but all use the declarative mode, which consists of interviewing individuals whose responses constitute the data (Daniel Coumon, 2016; p. 72). This technique will enable us to collect qualitative data by interviewing the different strata of the Goma community. This will include government officials, environmental experts, local businesses, NGOs, and so on. These interviews will provide valuable information on the challenges, needs, available resources and prospects for biogas production.
- **Observation technique**: Observation is a data collection technique that provides access to data that cannot be accessed through interviews or questionnaires (Fuincoise Chevalier and Sebastien Strenger, 2018; p. 94). Field observations will be essential to collect empirical data directly in Goma. Here, we will assess the biomass sources in place, visit the organic matter management facilities, and so on. These observations provided concrete information on local conditions and verified the initial hypotheses.

11.3.3.2.   Methods and techniques used in data analysis

Data analysis is a field from the world of statistics that aims to link different statistical data in order to classify, describe and analyse them succinctly (American Business School, 2022; p. 22).

A) Methods used to analyse the data
- **Experimental method**: is a scientific approach that consists of checking the validity of a hypothesis by means of repeated tests, during which the situational parameters are changed one by one in order to observe the effects induced by these changes (Pierre Grelley, 2012; p. 23). This method has been used to test biogas production at different scales, with the aim of validating our initial hypotheses. We will be setting up practical experiments to assess the feasibility of biogas production in the town of Goma. These experiments will enable us to observe and measure the results produced under real conditions, in order to gain a better understanding of the practical aspects of biogas production.
- **The analytical method**: is a research method that emerges from the scientific method and is used in the natural and social sciences for diagnosing problems and generating hypotheses that enable them to be solved. Using the analytical method, we were able to break down the biogas production process into its constituent elements and analyse the relationships between these elements. This helped us to better understand the technical and environmental issues related to biogas production in the city of Goma and to identify the key success factors for a high-quality feasibility study.

B) Techniques used in data analysis
- **Modelling and simulation techniques**: modelling thus makes it possible to project, in simulations, constructions or actions, to plan interventions and to organise the implementation

of the resources required for their realisation; thanks to simulations, it also has a preparatory function (J-M Der Maren, 2014; p. 239). These techniques were used to predict the performance of the biogas production system on the basis of the data collected. This included modelling the anaerobic fermentation process, simulating biogas production under different scenarios, projecting environmental benefits, etc. These techniques will help to assess predicted performance and identify areas for improvement.

11.3.3.3. Methods and techniques used to present and interpret the data

The presentation of the results of a scientific study is the first step in a process of publication and recognition (C. Bordenave, 2015; p. 1), and at its simplest, data interpretation involves examining the data collected through predefined processes in order to attribute meaning to this primary data and then draw a relevant conclusion (PIETRO MARZO, 2022; p. 2).

A) **Methods used to present and interpret the data**

❖ **Descriptive method**: refers to methods that describe the characteristics of the variables studied; thus it consists of observing and describing the behaviour, characteristics or conditions of a population or a particular phenomenon without manipulating any variable (Angelica Solomao, 2023; p. 45). It was used to draw up a complete portrait of the town of Goma in order to gain a better understanding of its geographical, political-administrative, economic, socio-cultural and environmental context. It also enabled me to describe the potential for biogas production in Goma.

❖ **Statistical method**: is an investigative procedure that requires all numerical data relating to a category of facts to facilitate their presentation and interpretation (GAWITZ M., cited by Desire Ngirumpatse, 2008; p. 9). It will be used to facilitate the collection of quantitative data on specific aspects of biogas production in the city of Goma. For example, data on available biomass resources, production costs, ... This method will enable us to properly process and analyse this data to provide measurable and objective results.

B) **Techniques used to present and interpret data**

❖ **Design technique**: is a creative, multidisciplinary and humanistic intellectual process, the aim of which is to address and provide solutions to everyday problems, large and small, linked to economic, social and environmental issues (AFD, 2022; p. 1).

II.2.4. Population and choice of sample

- **Study population:** Since a population is a finite or infinite subset of predefined elements on which observations can be made (M. GRAWITZ, cited by Rizou, 2017, p. 41), we have determined the population of the town of Goma as our study population.
- **Target population:** Our target population will include various sections of the population of the city of Goma, such as environmental experts, local authorities, students, other members of the community and companies interested in biogas.
- **Sample size:** As our target population is made up of various strata of the population of the city of Goma, the number of which is not well known, we proposed a simple random sample of 49 people, including 5 environmental officers, 5 leaders and local authorities, 5 students, 1 company (in its capacity as a legal entity) and 34 other members of the population of the city of Goma.

II.3.5. Data collection

Data collection is a systematic approach to gathering and measuring information from a variety of sources. This approach helps to obtain an overview of an area of interest (Louana Lelong, 2023; p. 2). This is why, methodically, we used the Empirical Method; the Documentary Technique; the Interview and Survey Technique; and the Observation

Technique.
Partial conclusion
In this chapter, we present the research methodology adopted for our feasibility study on biogas production in the city of Goma. We began by introducing the study environment, describing the geographical, environmental, social and economic characteristics of the city of Goma.

Next, we discussed the materials and tools used to carry out our experimental research, ensuring accurate and reliable data collection. We also detailed the experimental process, explaining the steps taken to collect the samples, carry out the measurements and conduct our biogas production study.

Chapter III
PRESENTATION OF RESULTS
This chapter, "Presentation of results", is a key stage in the study, where we aim to share the results obtained from our survey. In this chapter, we will highlight the objective of our survey and look at the interpretation and discussion of the results.

III.1. OBJECTIVE OF THE SURVEY
The aim of the survey is to assess the feasibility of biogas production in the town of Goma.

III.2.2. INTERPRETATION OF RESULTS

III.2.2.1. QUESTIONS ANCILLARY TO THE SURVEY
In this section, we present the results of the survey conducted as part of the feasibility study for biogas production in the town of Goma. The results below provide an overview of the data collected and the conclusions drawn from the survey:

III.2.2.1.1 Survey identity
Table IV. Age of respondents

Answers	Frequency	Percentage
15 to 20 years	1	2,0
21 to 30 years	24	49,0
31 to 40 years	9	18,4
41 to 50 years	12	24,5
Over 50s	3	6,1
Total	49	100

Source: Our field surveys in August 2023

From this table we can see that 24 people out of 49, or 49% of the surveys, are aged between 21 and 30. 12 people out of 49 (24.5%) are aged between 41 and 50. 9 people out of 49 (18.4%) are aged between 31 and 40. 3 people out of 49 (6.1%) are aged 50 and over. And finally, 1 person or 2% of respondents are aged between 15 and 20.

Table V. Civil status of the surveys

Answers	Frequency	Percentage
Single	26	53,1
Marie(e)	20	40,1
Widowed	2	4,1
Divorce	1	2,0
Total	49	100

Source: Our field surveys in August 2023

For the marital status variable, 26 people out of 49 (53.1% of surveys) are single, 20 people out of 49 (40.9% of surveys) are married, 2 people (4.3% of surveys) are widowed and 1 person out of 49 (1.4% of surveys) is divorced.

Table VI. Sex of respondents

Answers	Frequency	Percentage
Male	24	49,0
Female	25	51,0

| Total | 49 | 100 |

Source: Our field surveys in August 2023
In this table, we can see that 25 out of 49 people, or 51% of respondents, are women, while 24 out of 49, or 49% of respondents, are men.

Table VII. Level of education

Answers	Frequency	Percentage
Primary	5	10,2
Secondary	7	14,3
University	31	63,3
No level	6	12,2
Total	49	100

Source: Our field surveys in August 2023

This table shows that 31 people out of 49, or 63.3% of respondents, have a university education, while 7 people out of 49, or 14.3% of respondents, have a secondary education, 6 people out of 49, or 12.2% of respondents, have a primary education, and 5 people out of 49, or 10.2% of respondents, have no education.

Table VIII. Occupation of respondents

Answers	Frequency	Percentage
Commergant	10	20,4
Developer	3	6,1
Teacher/Trainer	9	18,4
Farmer	2	4,1
State agent/Local leader	13	26,5
Student	6	12,2
Contractor	4	8,2
Unemployed	2	4,1
Total	49	100

Source: Our field surveys in August 2023
The results of this table show the distribution of professional categories of our surveys. 13 people out of 49, i.e. 26.5% of respondents, are government employees and/or local leaders, 10 people out of 49, i.e. 20.4% of respondents, are shopkeepers, 9 people out of 49, i.e. 18.4% of respondents, are teachers and trainers. 6 out of 49 people (14.2%) surveyed were students 4 out of 49 people (10%) surveyed were entrepreneurs. 3 out of 49 people (6.1%) were developers. 2 out of 49 people (4.1%) were unemployed and 2 out of 49 people (4.1%) were farmers.

III.2.2.1.2. Perception of the surveys on biogas production in the city of Goma
Table IX. Public awareness of biogas production in the town of Goma

Question	Answers	Frequency	Percentage
Have you ever heard of biogas production in Goma?	Yes	48	98,0
	No	1	2,0

	Total		49	100

Source: Our field surveys in August 2023

According to the information gathered in this table, 48 out of 49 people (98% of respondents) had already heard of biogas production in the city of Goma. However, 1 person out of 49 (2% of respondents) had never heard of it.

Table X. In-depth knowledge of biogas production in the city of Goma

Question	Answers	Frequency	Percentage
If so, do you have in-depth knowledge of technology on biogas production?	Yes	2	4,1
	No	47	95,9
Total		49	100

Source: Our field surveys in August 2023

The results of this table show that 47 people out of 49 (95.9% of respondents) do not have in-depth knowledge of biogas production technology in Goma, while 2 people out of 49 (4.1% of respondents) do.

Table XI. Biogas production initiatives in the town of Goma

Question	Answers	Frequency	Percentage
Do you know of any existing biogas production initiatives in Goma?	Yes	10	20,5
	No	39	79,5
Total		49	100

Source: Our field surveys in August 2023

The results of this table show that 39 out of 49 people (79.5% of respondents) are not aware of these initiatives. And 10 people out of 49, i.e. 20.5% of respondents, said they were aware of them.

Table XII. Advantages of biogas production in the town of Goma

Question	Answers	Frequency	Percentage
What do you think are the advantages of biogas production in Goma?	Reducing the amount of organic waste	8	16,3
	Promoting renewable energy	11	22,4
	Reducing dependence on fossil fuels	9	18,4
	Job creation	21	42,9
Total		49	100

Source: Our field surveys in August 2023

According to this table, 21 people out of 49, or 42.9% of respondents, believe that biogas production in the town of Goma can help create jobs. 11 people out of 49 (22.4% of respondents) thought that it could promote renewable energy. 9 people out of 49, or 18.4% of respondents, thought that biogas production could reduce dependence on fossil fuels, while 8 people out of 49, or 16.3% of respondents, thought that it could reduce the amount of organic waste.

Tableau XIII. Participation rates in biogas production projects in Goma

Question	Answers	Frequency	Percentage
Would you be prepared to support or take part in biogas production projects in Goma?	Yes	34	69,4
	Perhaps, depending on the details of the project	15	30,6
Total		49	100

Source: Our field surveys in August 2023

The results of the survey indicate that 34 people out of 49, or 69.4% of respondents, said they were prepared to support or take part in biogas production initiatives in Goma, while a smaller proportion of 15 people out of 49, or 30.6% of respondents, were open to this possibility, subject to the details of the project.

Tableau XIV. Biogas: a substitute or replacement for the energy currently used in Goma

Question	Answers	Frequency	Percentage
What energy sources are currently used in Goma that could gradually be replaced or substituted by biogas?	The embers	26	53,1
	Firewood	4	8,2
	Oil (kerosene)	3	6,1
	Natural gas	13	26,5
	Solar energy	2	4,1
	Petrol/Oil/Diesel	1	2,0
Total		49	100

Source: Our field surveys in August 2023

According to the above results, the main sources of energy currently used in Goma and likely to be replaced or substituted by biogas are embers, according to 26 out of 49 people (53.1%) surveyed. Natural gas, according to 13 people out of 46, i.e. 26.5% of respondents, and petroleum (kerosene), according to 3 people out of 49, i.e. 6.1% of respondents. Firewood was chosen by 4 out of 49 people, i.e. 8.2% of respondents; solar energy was chosen by 2 out of 49 people, i.e. 4.1% of respondents. Finally, petrol/gas oil/diesel was chosen by 1 out of 49 people, i.e. 2% of respondents.

Table XV. Adoption of biogas use in the town of Goma

Question	Answers	Frequency	Percentage
Would it be possible for the people of Goma to adopt the use of biogas?	Yes	35	71,5
	Probably yes	13	26,5
	Not really	1	2,0
Total		49	100

Source : Our field surveys in August 20 |23

According to the results of this table, 35 people out of 49 (71.5% of respondents) believe that biogas can be adopted in the city of Goma. 13 other people out of 49, i.e. 26.5% of respondents, accept biogas with hesitation, while one person out of 49, i.e. 2% of respondents, considers that biogas cannot be adopted.

III.2.2.2. QUESTIONS RELATING TO THE PROBLEMATIC OF THE STUDY
A. Is it possible to produce biogas in the town of Goma?

After examining the possibility of producing biogas in the town of Goma, we concluded that it is practically feasible. Through extensive field research and the application of our knowledge in our test environment, we have determined that it is indeed possible to produce biogas in this region, provided that certain specific conditions and techniques are met. The resources required for this production are abundantly available, allowing them to be used for a variety of purposes.

By manually adapting certain equipment, we found that the majority of the population of Goma could produce and use biogas at their convenience. To put this hypothesis to the test, we carried out an experiment using a 2L bottle from 1 to 23 July 2023, with a ratio of 1:1 between the water and the fresh cow dung we used; this resulted in the production of biogas. Due to time and resource constraints, we carried out these trials on a small scale.

These results demonstrate the feasibility of biogas production in the town of
Goma, opening the door to promising prospects for its widespread use in the near future.

From 1er to 23 July 2023

Figure 4. Prototype utilisé lors des essaies
Figure 4. Prototype used for testing

Table XVI. Organic resources available in the town of Goma to be used to produce biogas

Question	Answers	Frequency	Percentage
What organic resources are available in Goma that could be used to produce biogas?	Food waste from urban markets, factories and households	28	57,1
	Agricultural waste (plant and/or animal residues)	14	28,6
	Aquatic waste from Lake Kivu or from fishing activities.	7	14,3
Total		49	100

Source: Our field surveys in August 2023

According to the results of this table, the organic resources available in the city of Goma to produce biogas are food waste from urban markets, factories and households, according to 28 out of 49 people (57.1%) surveyed, agricultural waste (plant and/or animal remains) for 14 out of 49 people (28.6%) surveyed, and aquatic waste from Lake Kivu or from fishing activities for 7 out of 49 people (14.3%) surveyed.

Table XVII. Obstacles or challenges encountered in implementing biogas production projects in Goma

Question	Answers	Frequency	Percentage
What might be the main obstacles or challenges in the implementation of projects to produce biogas in Goma?	Lack of awareness and knowledge about biogas in Goma	17	34,7
	Financial constraints for biogas infrastructures.	24	49,0
	Difficulties in managing and collecting organic waste in Goma	8	16,3
Total		49	100

Source: Our field surveys in August 2023

The table shows that 24 people out of 49, or 49% of respondents, consider that the financial constraints associated with biogas infrastructure are the main obstacles to implementing such projects in Goma. 17 people out of 49 (34.7%) believe that the lack of awareness and knowledge about biogas in Goma is also a major challenge. Finally, 8 out of 49 respondents (16.3%) pointed to the difficulties in managing and collecting organic waste in Goma as an obstacle to implementing biogas production projects.

Table XVIII. Strategies for setting up an efficient biogas production infrastructure in the city of Goma

Question	Answers	Frequency	Percentage
What strategies are needed to set up an efficient biogas production infrastructure in the city of Goma?	Collecting organic materials to ensure a constant supply of raw materials for biogas	10	20,5
	Using appropriate technologies to optimise biogas production in Goma	23	46,9
	Biogas awareness and education campaign in Goma	13	26,5
	Ensuring adequate access to storage and distribution infrastructure for biogas consumption	3	6,1
Total		49	100

Source: Our field surveys in August 2023

According to the results, four elements were identified as priorities for biogas production in the city of Goma. Firstly, according to 23 out of 49 respondents (46.9%), it is essential to use appropriate technologies to optimise biogas production. Secondly, according to 13 out of 49 respondents (26.5%), a biogas awareness and education campaign should be carried out in Goma, and the regular collection of organic materials should be ensured to guarantee a constant supply of raw materials for biogas production, according to 10 out of 49 respondents

(20.5%). Finally, 3 people out of 49 (6.1%) recommend adequate access to storage and distribution infrastructure to facilitate biogas consumption.

IIL2.2.3. Other questions
B. How can biogas be produced in Goma?

Table XIX. Biogas production process in the town of Goma

№	EXPERIMENTAL STEPS TAKEN TO PRODUCE THE BIOGAS IN THE CITY OF GOMA	INTERPRETATION
01.	Identifying sources of organic matter	The first step was to identify potential sources of organic material that could be used for biogas production. We found food waste from urban markets, factories and households, as well as agricultural waste and rubbish. We were particularly interested in cow dung because of the way it is used.
02.	Identification and collection of materials	Once the sources of organic matter had been identified, it was necessary to identify the equipment needed to produce the biogas. We assembled a tank, plastic bottles, piping (PVC and flexible hoses), valves, reducers, nipples, machons and fittings.
03.	Making a prototype	Once all the materials had been identified, it was time to build a prototype of the anaerobic digester. This involved building an airtight tank that would enable us to carry out anaerobic fermentation of the organic matter.
04.	Collecting OM	Once the prototype had been manufactured, it was important to collect organic matter from the sources identified. We were interested in cow dung because of its methanogenic bacterial content, and collected a sufficient quantity fresh.
05.	Loading the digester	Once collected, the cow dung, being our raw material, was loaded into our prototype at a ratio of 1:0, i.e. 0.75 litres of water and 0.75 kg of the MP. This makes a total of 1.5 litres of loading, leaving 0.5 for the gas collection space known as the biogas collector. The anaerobic digester gas collector is an essential part of the biogas production system, enabling the biogas produced to be collected and directed for efficient and safe use.
06.	Anaerobic fermentation	In the digester, the organic matter undergoes anaerobic fermentation, guided by microorganisms present in the anaerobic environment of the digester. During this

		process, the micro-organisms break down the organic matter into gases, mainly methane (CH4) and carbon dioxide (CO2). This process took us 23 days, from 1er to 23 July 2023.
07.	Gas collection	After the 4 biochemical stages, the gas is concentrated in the aerial part known as the biogas collector. We burnt this biogas to make sure it was real and stable.
08.	Monitoring and maintenance	Once the biogas has been collected, it is important to monitor the fermentation process on a regular basis, and to keep the digester in good working order. This can include activities such as measuring biogas production, monitoring temperature and pH, and cleaning and maintaining the digester.

Source: Our experimental studies; from April 2023 to August 2023.

These experimental steps represent a general process for biogas production in the city of Goma. It is important to note that the specific details and local conditions may vary depending on the design of the system, the resources available and the constraints specific to Goma.

C. What purpose would biogas production serve in the town of Goma?

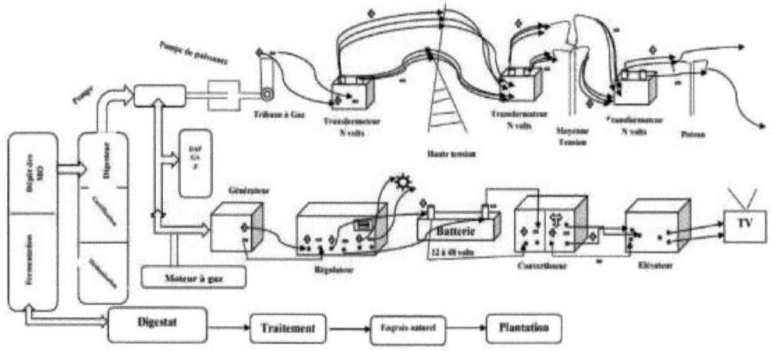

Figure 5. Diagram: Use of biogas

Source: Sekimonyo Shamavu Christian (2023; p. 42)
Interpreting the diagram :
This diagram shows 5 uses for biogas
1. **Production of high-voltage alternating current**
Here a large quantity of organic matter (of plant or animal origin) must be fermented in an anaerobic digester, connected to a high-pressure pump. As the organic matter is methanised, the pump ejects quantities of continuous-flow gas into impeller turbines, connected in series to produce direct current. To obtain alternating current, the anodes and cathodes are connected to a high-voltage transformer. The transformer carries the alternating current over high-voltage lines. This is followed by the transformation of the high-voltage current into low-voltage current and the transformation of the low-voltage current into medium-voltage current. This is where the connection comes in.
2. **Production of domestic alternating current**
The biogas produced by the digester will be fed into a gas generator that produces usable domestic electricity. This generator must be connected to a regulator to control the voltage exchange between the generator and the storage batteries. The latter are connected to a converter to supply a seamless electricity network.
3. **The use of biogas in gas engines**
BIOGAS can be used in gas engines for transport, or to process raw materials.
4. **Uses in the manufacture of digestate**
The residual organic matter in the digester serves as a natural fertiliser.
5. **Its use in cooking food**
Gas cylinders are suitable for storing biogas for cooking purposes. For this purpose, they must be connected to a heating plate;
111.2.3.   **DISCUSSION OF THE RESULTS**
111.2.3.1.   **Discussion on the possibility of producing biogas in the town of Goma**
In view of the applications we carried out in our test environment in July 2023 and the survey we carried out among the population of the town of Goma in August 2023 to verify our hypotheses, here is what we found:
After in-depth analysis and field research, we have come to the conclusion that biogas

production in the city of Goma is feasible in practice. Using our expertise and applying specific techniques, we have determined that this region does indeed have the capacity to produce biogas, provided certain conditions are met. The resources needed to produce biogas are widely available, offering many possibilities for use. Zhang, X., et al (2019, P9) studied the evaluation of biogas production from food waste in a large Chinese city. Their analysis showed that the implementation of anaerobic digestion systems in cities could not only reduce waste, but also provide a clean and renewable energy source for the urban community.

This is how we put forward our main hypothesis, which is that the biogas production process could be possible in the town of Goma using an anaerobic digestion plant that converts organic waste into biogas and digestate.

111.2.3.3. Discussion on the resources available in Goma for biogas production

With regard to the resources available in Goma to produce biogas, as shown in Table **XVIIIXX**, 28 out of 49 people (57.1%) surveyed used food waste from urban markets, factories and local kitchens, 14 out of 49 people (28.6%) surveyed used agricultural waste (plant and/or animal remains) and 7 out of 49 people (14.3%) surveyed used aquatic waste from Lake Kivu or fishing activities. These results take into account half of our 3rd hypothesis, where we spoke only of plant and animal remains. Here we have 57.1% proof that food waste from urban markets, factories and local kitchens is more available, to which we can add O.M. (organic matter) from Lake Kivu or fishing activities.

111.2.3.4. Discussion on the obstacles to biogas production in the city of Goma

With regard to the main obstacles or challenges that may be encountered in implementing biogas production projects in Goma, Table XIX shows that the majority of respondents (49%) consider that the financial constraints associated with biogas infrastructure are the main obstacles to implementing such projects in Goma. Next, 34.7% of respondents felt that the lack of awareness and knowledge about biogas in Goma was also a major challenge. Finally, 16.3% of respondents identified the difficulties of managing and collecting organic waste in Goma as an obstacle. These results confirm our 2eme hypothesis about bottlenecks in setting up a biogas production project in the city of Goma.

111.2.3.5. Discussion of strategies to be implemented

The strategies to be taken into account to set up an efficient biogas production infrastructure in the city of Goma, we can read in table XX that it is essential to use adapted technologies in order to optimise biogas production, which represents 46.9% of the responses. Next, 26.5% of respondents said that an awareness and knowledge campaign on biogas should be carried out in Goma, and that organic matter should be collected regularly to guarantee a constant supply of raw materials for biogas production, according to 10 out of 49 respondents (20.5%). Finally, adequate access to storage and distribution infrastructure is necessary to enable biogas consumption, which accounts for 12.3% of responses. These responses confirm our last hypothesis.

After in-depth analysis and field research, we have come to the conclusion that biogas production in the town of Goma is practically feasible. Using our expertise and applying specific techniques, we have determined that the town of Goma does indeed have the capacity to produce biogas, provided certain conditions are met. The resources needed to produce biogas are widely available, offering a wide range of possible uses.

III.4. Limits and constraints

During our study entitled "Feasibility study of biogas production in the city of Goma, from

April to August 2023", we encountered various constraints that hampered our progress. These included insufficient material, financial and human resources, as well as time and space constraints.

From a material point of view, we had to make do with the limited resources available to us to carry out our study on the trial site. The financial means were also insufficient, which forced us to carry out small-scale trials with a limited number of available materials.

The time available was also a major constraint. It was not enough to carry out large-scale experiments and collect in-depth data in the field. This time constraint limited the scope of our study and required careful planning and organisation to optimise the time available.

In addition, the lack of space was a major challenge. The space available was not adequate to carry out trials efficiently and satisfactorily. This limited our ability to carry out experiments in line with our requirements. We had to deal with this constraint by optimising the space available and adapting our research methods.

Despite these constraints, we have completed our survey to the best of our ability. It is important to note that these constraints can be overcome with adequate resources and better planning. These limitations also highlight future needs in terms of resources and infrastructure to promote research and development in biogas production in Goma.

Partial conclusion

In this chapter, we have demonstrated the feasibility of producing biogas in the city of Goma, taking certain factors and techniques into account. It is clear that biogas production in Goma is practically feasible. Our studies focused on the resources available in Goma for biogas production, as shown in Table XXI. The results showed that food waste from urban markets, factories and households is the most abundant, followed by agricultural and aquatic waste from Lake Kivu or from fishing activities. These results confirmed half of our 2eme hypothesis, in which we had mentioned only plant and animal remains. Food waste from urban markets was shown to account for 57.1% of available resources, demonstrating the high potential for biogas production in Goma.

A study carried out by Robert H. Beach in his article "Biogas Production from Solid Waste in Developing Countries" (2010, pp. 1784) argues that anaerobic digestion systems can effectively exploit this waste to produce biogas.

However, obstacles and challenges must be overcome in order to implement biogas production projects in Goma, as shown in Table XIX, which shows that the majority of respondents (49%) consider that financial constraints linked to biogas infrastructure are the main obstacles to implementing such projects in Goma. Next, 34.7% of respondents felt that the lack of awareness and knowledge about biogas in Goma was also a major challenge. Finally, 16.3% of respondents identified the difficulties of managing and collecting organic waste in Goma as an obstacle. These results corroborate our third hypothesis on the obstacles to biogas production in Goma.

On the other hand, it is important to note that the general public is insufficiently aware of the benefits of renewable energies and their potential to meet the energy challenges of each country, according to S. Niang et al. in their 2017 article (p. 25).

To establish an efficient biogas production infrastructure in Goma, various strategies were proposed, as shown in Table XX. It is essential to use appropriate technologies to optimise biogas production, which accounted for 46.9% of responses. Next, 26.5% of respondents said that an awareness and knowledge campaign on biogas should be carried out in Goma. Regular collection of organic materials should be ensured to guarantee a constant supply of raw materials for biogas production, according to 10 out of 49 respondents (20.5%). Finally, adequate access to storage and distribution infrastructure is needed to enable biogas consumption, which accounted for 12.3% of responses. These strategies support our fourth hypothesis. On the one hand, researchers have described a standardised technology for planning and selecting a technology adapted to the specific needs of a particular context to make biogas production viable (Amon et al., 5th ed., 2010, p. 17). On the other hand, in many developing countries there is a lack of infrastructure for disseminating information to technicians, policy-makers and potential users. Ongoing qualitative and quantitative evaluation of activities is needed to improve the technology and adapt its application to each country (Marchaim et al., 1994, p. 8).

Finally, our study has demonstrated the feasibility of biogas production in Goma, taking into account the resources available. However, challenges such as financial constraints, lack of awareness and waste management difficulties need to be overcome for the successful implementation of biogas production projects. The proposed strategies, including the use of adapted technologies, awareness-raising campaigns and regular collection of organic materials, are essential for an effective biogas production infrastructure in Goma. It is important to continue to innovate and monitor technological progress in order to maximise the potential of biogas as a renewable energy source in the city of Goma and meet the energy challenges in a sustainable way.

Chapter IV
PROPOSED STRATEGIES
III.0. INTRODUCTION
In this IVeme chapter, we propose strategies for biogas production in Goma. These strategies are designed to meet the challenges identified, such as raising awareness, collecting organic matter, using appropriate technologies and gaining access to storage and distribution infrastructure.

Using Harold Lasswell's '5W and 1H' model cited by Campesato (2023, p. 197), we will be able to develop effective strategies to address these key issues. This will enable us to set up sustainable biogas production projects that will benefit the Goma community and the environment.

III.1. FORMER STRATEGIES
To overcome the difficulties associated with biogas production, solutions have been proposed by authors and experts in the field. Thus, we present some old strategies that have been suggested in the biogas sector, in response to the challenges identified:

1. **Wendy LAPERRIERE (2017);** in her thesis on *"Study of the limits of biogas digesters in flexible operation"*, proposed: control to maintain process stability and maximise biogas production and monitoring of certain parameters such as substrate concentration, temperature and pH must.

2. **Yaqin Zheng, Jeng Xu et al (2015)**; in their study entitled "Treatment techniques for enhancing biogas production from food waste: A review", proposed the production of biogas from food waste; the optimisation of anaerobic digestion parameters and the use of advanced technologies, such as two-phase methanisation or sequential digestion.

3. **Liebetrau, Krierg et al (2013)**; in their practical guide *"Biogas Handbook: Science, Production and Applications"* have proposed several strategies for optimising the biogas production process, including the use of mixed cultures, optimising substrate mixing and improving the management of end products.

4. **RECORD (2022);** in its report *"Stockage et pretreatments des intrants avant alimentation de digesteurs de methanisation - Etat des connaissances et recommandations"* (Storage and pretreatment of inputs before feeding methanisation digesters - State of knowledge and recommendations), for the improvement of the biogas sector, it is suggested that agricultural inputs, R&D (Research and Development) efforts and the extended analysis of the impacts of processes from an environmental, energy and health point of view be brought together.

5. **Laurent Lardon and Bernard Ribeiro (2012)**, in their book *"Biomethanisation"*, propose the selective collection of waste; the creation of waste treatment centres; the use of advanced technologies to maximise the potential for converting organic waste into energy; and finally the involvement of the community to encourage the reduction of waste at source and greater environmental responsibility.

All the proposed strategies listed above were appropriate and applicable in different environments depending on the challenges identified. For our part, we propose in this section strategies related to the local context of the town of Goma, our study environment, where we had identified certain obstacles to achieving the production and adoption of biogas.

III.2. NEW STRATEGIES
In this sub-section, we present new strategies for making biogas production a reality to the economic and environmental standards of a majority in the city of Goma.

The major challenges to the effective implementation of a production project in the city of Goma include the following:
- Lack of awareness and knowledge about biogas in Goma
- Financial constraints for biogas infrastructures.
- Difficulties in managing and collecting organic waste in Goma

These challenges call for impeccable strategies if biogas production is to be effective and appropriate in the city of Goma. With this in mind, a number of strategies have been proposed and will be developed in this section. These strategies include
- A biogas awareness campaign in Goma;
- Collecting organic materials to ensure a constant supply of raw materials for biogas;
- Using appropriate technologies to optimise biogas production in Goma and the region
- Ensuring adequate access to storage and distribution infrastructures for biogas consumption.

III.2.1. Relationships between the strategic axes

The figure below shows how the specific strategies are linked around the main objective to interact in such a way that we can
reach it.

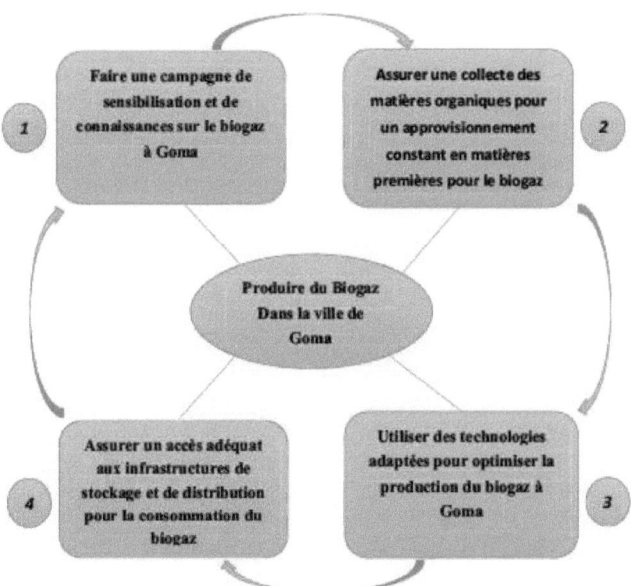

Producing biogas in the city of Goma
1 A biogas awareness and education campaign in Goma
2 Collecting organic matter to ensure a constant supply of
raw materials for biogas
3 Ensuring adequate access to storage and distribution infrastructure for biogas consumption
4 Using appropriate technologies to optimise biogas production in Goma
Figure 6. Logical model of strategic axes

III.2.2. Strategy development and implementation

These strategies are designed to promote the use of biogas in Goma. The first step is to raise awareness and educate the population about the advantages and uses of biogas. This will help to create demand and interest among residents in this alternative energy source.

The second strategy is to ensure the efficient collection of organic materials, which will be used as raw materials for biogas production. This ensures a constant supply of fuel and limits the obstacles to production.

The third strategy focuses on the use of appropriate technologies to optimise biogas production. This maximises the quantity and quality of the biogas produced, ensuring efficient use of available resources.

Finally, to encourage the use of biogas, it is essential to put in place adequate storage and distribution infrastructures. This ensures that the biogas produced is easily accessible to those who need it, thereby encouraging its uptake.

All these strategies are aimed at raising awareness and efficiently collecting, producing and distributing biogas in Goma, thereby fostering the development of a sustainable source of energy for the community.

III.2.2.1. Conduct an awareness and knowledge campaign on biogas in Goma.

By developing this strategy according to the 5Ws and 1H, it will be possible to implement an awareness and knowledge campaign on biogas in Goma, which will reach a wide audience and encourage wider adoption of this renewable energy. The strategic approach is to set up an awareness and knowledge campaign on biogas in Goma, using Harold Lasswell's five Ws (Who, What, When, Where, Why) and H (How) to guide the communication approach.

Table XXII. Strategic development of the biogas awareness and knowledge campaign in Goma

Who?	What?	When?	Where?	Why?	How?
1000 households in the city of Goma	The campaign aims to raise awareness of the benefits and applications of biogas. This includes promoting the use of biogas as a clean, renewable alternative to fossil fuels, reducing greenhouse gas emissions, managing organic waste sustainably and creating local jobs. The campaign should also provide information on biogas production processes, the equipment required and safety standards.	From January to June 2024 (6 months)	The campaign will cover the whole of Goma, using appropriate communication channels to reach different audiences. These will include information sessions in schools, neighbourhood meetings, conferences in local institutions, billboards in strategic locations, radio and television broadcasts, and online promotions on social networks.	The campaign aims to raise awareness among the people of Goma of the ecological, economic and social benefits of using biogas. By understanding why it is important to adopt this renewable energy, people will be more likely to take steps to do so.	The campaign can use a multi-channel approach to reach a wide audience. This can include oral presentations, educational videos, brochures, informative posters, viral videos on social networks and practical workshops on biogas production. Resource people, such as biogas experts and testimonials from those who have already adopted this energy, can also be involved to reinforce the message and promote wider adoption.

By developing this strategic scion the 5W and 1H, it will be possible to implement an awareness campaign and

of knowledge about biogas in Goma, which will reach a wide audience and encourage wider adoption of this renewable energy.

III.2.2.2. Collecting organic matter to ensure a constant supply of organic matter for biogas.

The strategic aims to ensure the regular collection of organic materials to guarantee a constant supply of the raw materials (RMs) needed to produce biogas. To develop this strategic approach, we follow Harold LasswelL's model of the five Ws (Who, What, Where, When, Why) and the H (How).

Table XXIII. Development of the strategic collection of organic matter to ensure a constant supply of organic matter for biogas.

Who?	What?	When?	Where?	Why?	How?
4 associations (Umoja Ni Nguvu, Kunafazi Usafi, Umoja Ni Nguvu, Mawe) collecting waste in the city of Goma.	Define the types of organic materials to be collected. These will include food waste from households, garden waste, biomass from the agri-food industry and other organic sources available in Goma.	From July to August 2024 (2 months)	The site will include specific collection points in residential areas (Lac Vert and Mugunga).	Ensuring regular collection of organic materials to guarantee a constant supply of the raw materials (PM) needed to produce biogas	The use of special collection bins in residential areas, the installation of collection points on farms and/or the development of collaboration with supermarkets and restaurants to recover their food waste.

In developing this strategic approach, it is important to raise awareness among the population of Goma of the advantages of biogas and the need to use it.

establish partnerships with the relevant players.

IIL2.2.2. Using adapted technologies to optimise biogas production in Goma

The strategic approach consists of using appropriate technologies to optimise biogas production in Goma, taking into account the elements set out in the table below:

Table XXIV. Development of the strategic use of adapted technologies to optimise biogas production in Goma

Who?	What ?	When?	Where?	Why?	How?
Government authorities, non-governmental organisations (NGOs), private companies and the local population. They will work together to implement this strategic.	The use of adapted technologies to improve biogas production in Goma. This includes installing biodigesters, filtration systems and methods for sorting organic waste using locally available equipment.	Throughout the project period.	This strategic alliance will be set up in Goma, a city in the Democratic Republic of Congo in the province of North Kivu.	Use adapted technologies (cheaper, locally accessible equipment) to optimise biogas production in Goma	Assess the population's energy needs and material resources. Draw up a plan for the installation of biodigesters adapted to local needs. Regularly monitor the effectiveness of the strategic approach and make any necessary adjustments.

By implementing this strategic approach, Goma will be able to optimise its biogas production and reduce organic waste,

promoting the circular economy and improving its supply of clean energy.

III.2.2.2. Ensure adequate access to storage and distribution infrastructures for biogas consumption

The strategic approach consists of ensuring adequate access to storage and distribution infrastructures for the consumption of biogas by contributing the elements contained in the table below:

Table XXV. Strategic development of adequate access to storage and distribution infrastructures for biogas consumption

Who?	What?	When?	Where?	Why?	How?
The key players involved in this strategic approach are the local authorities, technology and equipment suppliers, local businesses and the community.	The aim is to set up biogas storage and distribution infrastructures so that biogas can be consumed on a large scale.	From September to October 2024.	This strategic approach is being implemented in Goma, a town in the province of North Kivu in the Democratic Republic of Congo.	Ensuring adequate access to storage and distribution infrastructure to encourage biogas consumption.	Assessing the needs in terms of biogas storage and distribution in Goma. Draw up a development plan for the necessary infrastructure, taking into account the city's specific characteristics and technical and environmental constraints. Mobilise financial and technical resources to implement this plan, by promoting public-private partnerships and seeking funding from national and international organisations.

By setting up adequate storage and distribution infrastructures, this strategic initiative will ensure easy access to biogas for consumption in Goma, thereby contributing to the sustainable development of the region.

GENERAL CONCLUSION

Biogas represents a promising and sustainable source of energy for underdeveloped countries in general and the DRC in particular. In these contexts, where access to energy is often limited, biogas offers many advantages as an energy solution.

In this study, which focuses on *"Biogas production in the city of Goma: feasibility study from April to August 2023"*, our overall objective is to study the feasibility of producing biogas in the city of Goma.

To achieve this objective, we started from a base based on the following questions:
1. Is it possible to produce biogas in Goma?
2. What are the most important sources of organic matter in Goma for producing biogas?
3. What obstacles are there to biogas production in Goma?
4. What strategies are needed to set up an efficient biogas production infrastructure in the city of Goma?

In response to these concerns, we have formulated the following hypotheses:
1. The biogas production process could be possible in the town of Goma using an anaerobic digestion plant that converts organic waste into biogas and digestate.
2. Biodegradable waste from households, pigsties, barns, henhouses and mini-industries are the main sources of PM for biogas production in the city of Goma.
3. The lack of information and of material, financial and technical or technological resources would appear to be one of the major difficulties encountered in producing biogas in the town of Goma.
4. A biogas awareness and education campaign, the use of simple technology with locally available equipment, ensuring regular collection of organic matter and adequate access to storage and distribution infrastructure to facilitate biogas consumption in the city of Goma would be the strategies to be adapted to achieve biogas production in the city of Goma.

In order to verify our hypotheses, we successively used three methods: descriptive, statistical and experimental. These methods were supported by a number of techniques, including documentary techniques, interview and survey techniques, observation techniques and modelling and simulation techniques.

After an in-depth analysis and field research, we can conclude that biogas production in the city of Goma is practically feasible. The resources available in Goma to produce biogas were studied, as shown in Table **XVIII**. The results show that food waste from urban markets, factories and local kitchens is the most available, followed by agricultural waste and aquatic waste from Lake Kivu or fishing activities. This diversity of organic waste sources confirms the high potential for biogas production in Goma.

However, obstacles and challenges need to be overcome in order to implement biogas projects in Goma, as shown in Table XIX. The main constraints identified are financial constraints related to biogas infrastructure, lack of awareness and knowledge about biogas, and difficulties in managing and collecting organic waste. These results confirm our initial hypotheses on the obstacles to setting up biogas production projects in Goma.

To set up an efficient biogas production infrastructure in Goma, various strategies have been identified, as shown in Table XX. It is essential to use adapted technologies to optimise biogas production, to conduct an awareness and knowledge campaign on biogas, to ensure regular collection of organic matter and to guarantee adequate access to storage and distribution infrastructures. These strategies confirm our initial hypotheses.

In summary, our study shows that biogas production from available resources in Goma is feasible and has great potential. However, efforts need to be made to overcome financial barriers, raise awareness and improve waste management. By adopting the right strategies, Goma can put in place an efficient biogas production infrastructure that will contribute both to the production of clean energy and to the sustainable management of organic waste.

Bibliography

I. Dictionary
1. Le petit Larousse illustre, 2002
2. The Illustrated Robert, 2021

II. General works
1. ACF, (2011), *Construction d'un biodigesteur biogaz,* Toronto, Bathurst, 5 p.
2. AFG, (2018), *Gaz renouvelables : les leviers de développement,* Paris, EDF, 20 p.
3. AILE, (2022), *Substrates,* Marseille, Bioco, 3 p.
4. Almalowi, S.J. and Jobran, F.A. (2022) *Experimental Study of Design and Implementation of Small-Scale Biogas Digest System.* Energy and Power Engineering, 14, 156165.
5. Amadou Mamoudou et al (2019), Attitude de l'Afrique centrale face aux energies renouvelables, Yaounde, IFRIKIYA, 88 p.
6. Atee Club, (2014), *Le biogaz, une énergie renouvelable multiforme, strategique dans la transition,* Paris, Atee, 72 p.
7. BAUTISTA ANGELI J-M., (2019) *Etude de faisabilité de la micro-methanisation par codigestion a l'echelle des quartiers,* 202 p.
8. Beth Doerr and Nate Lehmkuhl, (2008), *Biodigesters a Methane,* Ottawa, Echo, 7 p.
9. BRIMO.A. (1971), <u>Methodes des sciences sociales,</u> ed. Montchrestien, Paris.
10. Bukanga and Sekimonyo, (2021), Notes de cours de probabilites statistiques, unpublished, G3 Petrole Gaz et Energies renouvelables/UNIKIN, 45 p.
11. Cedric Philbert (2018), *biogas, renewable energy and sustainable development,* Paris, IFRI, 52 p.
12. CGES, (2017), *urban development project,* Kinshasa.
13. CGES, (2017), DRC; mini-green grid programme, New-York, BM, 144 p.
14. Christian SEKIMONYO (2023), *problematique energetique : transition energetique,* SARRBRUCK, Editions Universitaires Europeennes, p. 52
15. Coba H. O. et al (2018), *Renewable energies in Latin America,* Quito, El conejo, 100 p.
16. DAYANA A, (2004), *Alternative energies in the Great Lakes countries,* Goma, CIRGL, 32 p.
17. Dupont Jeannine, (2020), *leprocede de la digestion anaerobie,* Quebec, Hamac, 38 p.
18. Fernandez-Polanco at al, (2000), *Fuel,* 83, 195-204
19. Filde ADEOSSI, (2013), *Caracterisation du biogazproduit a partir des substrats Bovins et porcins dans la region du centre du Burkina Faso,* Ouagadougou, inedit, Centre Commun De Recherche Eau-Climat, 54 p.
20. Frederik Belanger (2009), *Etude de faisabilité techno-economique et sociopolitique d'un projet regional de methanisation de lisier de porc en codigestion,* Quebec, Canada, SHERBROOKE, 136 p.
21. GrDF, (2019), *Methanisation agricole : Retour d'experience sur l'appropriation locale des sites en injection,* Paris, GRDF, 56 p.
22. Heidarzadeh, Younesi et al. (2017), *Biogas production from municipal solid waste: A review on process monitoring,* Cologne, IRJET, 102 p.
23. KAMPEMPE BUSILI Justin, (2014), Notes de cours de statistiques referentielles, inedit, G2, Mathematiques info/UNIKIN, 78 p.
24. Kampempe and Sekimonyo, (2022), Notes de cours Recherches Operationnelles,

unpublished, G3 Polytechnique/UNIKIN, 54 p.
25. Kretschmer, R., Gerbig, C., Karstens, U., & Koch, F-T. (2012), 'Error characterization of CO_2 vertical mixing in the atmospheric transport model WRF-VPRM. *Atmospheric Chemistry and Physics"*, *12*(5), 2441-2458.
26. Laurent Lardon and Bernard Ribeiro (2012), **Biomethanisation,** Ottawa, Ontario, 26 p.
27. Laurent Lardon and Bernard Ribeiro (2012), Biomethanisation, Paris, Gallimard, 558 p.
28. Liebetrau, Krierg et al (2013), *Biogas Handbook: Science, Production and Applications,* Cambridge, WPL, 500 p.
29. Liebetrau, Krierg et al (2013), *Biogas Herdbook: Science, Production and Applications,* Berlin, Metropol Verlag, 44 p.
30. Marchaim at al, (1994), "Biogas production processes for the development of sustainable technologies", FAO, Rome, 230 p.
31. Mario A. et al. (2019), *The industry of the future*, Quebec, UNISI, 10 p.
32. Muhammad Arslan et al, (2020), *Asia outlook Energy,* Gaza, Al-mansour, 38 p.
33. Ndekhedehe Efiong Edet et al, (2017), *de déchets au Gaz,* Douala, Doualatour, 76 p.
34. P. BUFFIERE, R. BAYARD et al (2009), *freins et developpements de la filiere biogaz : Les besoins en recherche et developpement,* Paris, RECORD, 148 p.
35. Paul Njogu, Robert Kinyua et al (2015), the issue raised in their article *Biogas production using water hyacinth (Eicchornia crassipes) for power generation in Kenya,* Nairobi, URPA
36. Rainer Janssen, Dominik Rutz et al (2010), *Bioenergy for sustainable development in africa - environmental and social aspects,* Munich, Germany, WIP, 9 p.
37. S. Kalloum, M. Khelaf et al, (2021), "*Influence du pH sur la production du biogaz a partir des déchets managers",* Alger, El-Malakia, 22 p.
38. S. Niang at al, (2017), *Energies renouvelables, transition energetique et enjeux climatiques en droit africain,* Pretoria, unpublished, 54 p.
39. Sama, H. & Tiabri Thiombiano, S., (2018), *Le biogaz a des fins domestiques*, Paris, Prisme, 8 p.
40. Samuel Gikundu et al, (2018), *Food waste for a green economy*, Nairobi, unpublished, 122 p.
41. TCHA-THOM Maglwa, (2019), *Recherche d'une filiere durable pour la methanisation des déchets de fruits et d'abattoirs du Togo : Evaluation du potentiel agronomique des digestat sur les sols de la Region de la Kara*", Lome, Michca, 204 p.
42. Thomas, J. (2020) *A Methodological Outlook on Bioplastics from Renewable Resources. Open Journal of Polymer Chemistry*, 10, 21-47. doi: 10.4236/ojpchem.2020.102002.
43. Wendy Laperriere, (2017), *Evaluation des limites d'un digesteur biogaz pour une utilisation flexible dans un reseau local de production d'energie,* Universite de Montpelier, Montpelier, 216 p.
44. Zheng, Y., Xu, J., et al. (2015), *Treatment techniques for enhancing biogas production from food waste: A review,* Lianui, HongFei, 266 p.

III. Articles and reviews
1. Robert H. Beach (2010) "Biogas Production from Solid Waste in Developing Countries" in Energy review, Cambridge, Cambridge University Press, 1780-1802 pp.
2. Twizerimana, M., M'Arimi, at al. (2021) **Anaerobic Digestion of Cotton Yarn Wastes for Biogas Production: Feasibility of Using Sawdust to Control Digester Temperature at**

Room Temperature. Open Access Library Journal, 8: e7654.
3. Hammad, E.I., Al-Agha, M.R. and El-Nahhal, Y. (2019) *Influence of Biogas Production on Bioremediation of Animal Manures*. American Journal of Analytical Chemistry, 10, 1-8.

IV. Report and other documents

1. Amon at al, (2010), **Benchmarking et sélection des technologies de pyrolyse et de gazeification adaptes a la valorisation des CSR et du Bois-B sous forme du gaz**. [Technical Report] Imt mines-albi
2. Emilien Dubiez, Laurent Gazull at al (2020), Rapport d'etude de la consommation en energies de production des usagers productifs de la ville de Goma, Goma, CIRAD, 40 p.
3. Emilien Dubiez, Laurent Gazull at al (2020), Rapport d'etude de la filiere bois-energie de la ville de Goma, op. cit, 60 p.
4. Mairie de Goma (2018) Rapport annuel de la mairie de Goma : *Division de l'economie*
5. Mairie de Goma (2020), *population census report for the city of Goma.*
6. Mairie de Goma, (2018) *annual report.*
7. Mairie de Goma, (2020) *annual report.*
8. PASA-NK, (2016), *Final design report,* GOMA.
9. UNDP (2013), *Sustainable Energy for All by 2030,* New-York, UN, 82 p.
10. Report, Provincial Governorate of North Kivu, 2006
11. **RECORD (2022)**, "*Stockage et pretraitements des intrants avant alimentation de digesteurs de methanisation - Etat des connaissances et recommandations*", Paris, 152 p.

V. Unpublished works

1. Coarita Frenandez, (2021), *Pretraitement des déchets agricoles pour l'optimisation de leur valorisation par methanisation,* Environnement et Societe, inedit, Universite de Lyon, 222 p.
2. Faballa NDIAYE, (1996), *developpement du biogaz par l'elevage : sites delta du fleuve Senegal et haute Casamance,* "doctoral thesis", unpublished, Dakar, E.I.S.M.V., 140 p.
3. IGUENANE RABAH, (2012), *Systeme d "alimentation energetique d "une ferme par combinaison digesteur anaerobie et eolienne*, Tizi ouzou, inedit, Universite Mouloud Mammeri, 94 p.
4. Joseph Kitaganya (2023), Notes de cours de philosophie de l'art, unpublished, ISAM-Goma,
5. Kanambe JUAKALY, (2012), *L'emergence du M23 au Nord-Kivu et son impact sur la situation socio-economique des habitants de la ville de Goma,* Goma, inedit, ISDR-GL
6. Kitaganya and Sekimonyo, (2023), *Notes de cours de MRS*, inedit, ISDR-GL, 62 p.
7. Nixon BIKA NTAMIRABALI, (2015), *Determinants de la consommation des energies renouvelables a Goma, cas de l'energie solaire,* Goma, inedit, UNIGOM, 78 p.
8. Nzibonera Bayongwa D. (2014), *la gestion rationnelle de la peche aux filets maillant sur le lac Kivu en republique democratique du Congo: strategies pour une peche durable et rentable,* unpublished, Distance Production House University, 156 p.
9. Pauline CAMPESATO, (2023), *La communication responsable : enjeux sociaux, ecologiques et éthiques* (doctoral thesis), Nice, inedit, Universite Cote D'Azur, 548 p.
10. Vincent NORDMANN (2013), *Characterisation and impact of different fractions d'une biomasse lignocellulosique pour ameliorer les pretraitements favorisant sa methanisation,* (doctoral thesis), inedit, Universite Bordeaux 1, Bordeaux, 214 p.
11. Yusuf MUHINDO, (2021), *L'elevage des poules pondeuses et son impact sur la vie economique dans la ville de Goma*, Goma, Inedit, ISDR-Goma

Webography
1. Φ www.water.gov.ma; anaerobic digestion, consulted on 12 June 2023 at 23:02'.
2. Φ https://fr.wiktionary.org/wiki/biogaz, accessed on 13 June 2023 at 00:02'.
3. Φ https://www.ifpenergiesnouvelles.fr; renewable energies ; consulted on 13 April 2023 at 00h03 min

Appendix
SURVEY QUESTIONNAIRE

As part of our end-of-study work in rural development, specialising in the environment and sustainable development, at the Institut Superieur de Developpement Rural des Grands Lacs (ISDR-GL), we are carrying out research into the feasibility of biogas production in the town of Goma, from April to August 2023. We would be grateful if you could answer our questions in complete anonymity and confidentiality.

Please tick the answer that corresponds to your choice for multiple-choice questions.

Thank you in advance.

A. IDENTITY OF THE SURVEY (E)
1. Age range

a) 10 to 20 years	c) 31 to 40 years	e) Over 50	J
b) 21 to 30 years	11 d) 41 to 50 years	11	

2. Civil status

a) Single	(c) Widowed	: i
b) Mary	d) Divorce(e)	□

3. Type of survey

a) Male	[b) Female

4. Level of education

a) Primary	I c) Academic	
b) Secondary	(d) No level	[

5. Profession

a) Commergant		c) Farmer	[
b) Teacher	d) Civil servant		

B. *QUESTIONS PROPER*

6) Have you ever heard of biogas production in the town of Goma?
a) Yes   b) No

7) If so, do you have in-depth knowledge of biogas production technology?
a) Yes   b) No

8) Would you be interested in learning more about biogas production and its environmental benefits?
a) Very interested
b) Moderately interested
c) Not at all interested

9) Is it possible to produce biogas in Goma?
a) Yes
b) No

10) Do you know of any existing biogas production initiatives in Goma?
a) Yes
b) No

11) In your opinion, what are the advantages of biogas production in the city of Goma?
a) Reducing organic waste
b) Production of renewable energy
c) Reducing dependence on fossil fuels [

12) What do you see as the main obstacles to setting up biogas production in Goma?

a) Lack of knowledge and awareness
b) Lack of financial resources
c) Lack of support and government policies

13) Would you be prepared to support or take part in biogas production projects in Goma?
a) Yes b   ) No
c) Perhaps, depending on the details of the project [?

14) What strategies are needed to set up an efficient biogas production infrastructure in the city of Goma?
a) Collecting organic materials to ensure a constant supply of raw materials for biogas.
b) Using appropriate technologies to optimise biogas production in Goma.
c) Ensuring adequate access to storage and distribution infrastructures for biogas consumption.

15) Biogas awareness and knowledge campaign

16) What are the main obstacles or challenges to implementing biogas production projects in Goma?
a) Insufficient awareness and knowledge of biogas in Goma.
b) Financial constraints for biogas infrastructures.
c) The difficulties of managing and collecting organic waste in Goma.

17) What organic resources are available in Goma that could be used to produce biogas?
a) Food waste from urban markets and local kitchens.
b)
c) Agricultural waste (plant and/or animal residues)
d) Aquatic waste from Lake Kivu or from fishing activities.

18) What are the potential environmental and economic benefits of biogas production in Goma?
a. Reducing greenhouse gas emissions by producing clean energy.
b. Reducing dependence on imported fossil fuels
c. Job creation
d. Making the city healthy

19) Would it be possible for the people of Goma to adopt the use of biogas?
a) Yes
b) No

20) What energy resources are currently being used in Goma that could gradually be replaced or substituted by biogas?
a) The ember ;
b) Firewood ;
c) Oil (kerosene)
d) Natural gas

Administrative map of the city of Goma

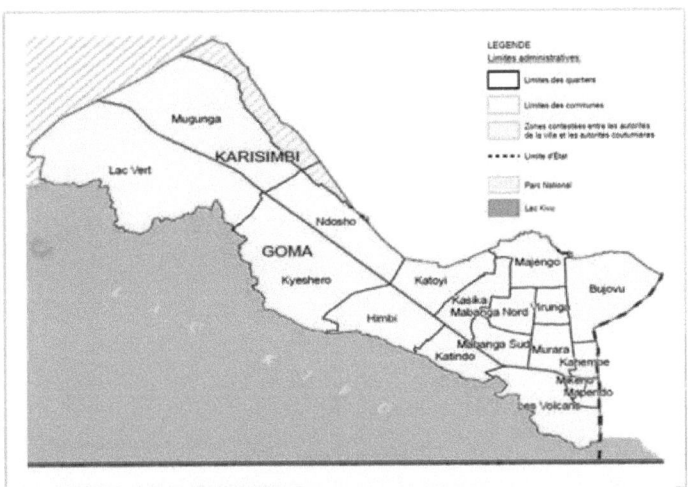

Figure 7. Carte Administrative de la ville de Goma
Figure 7. Administrative map of the city of Goma

Source: INS (2015; p. 4)

I want morebooks!

Buy your books fast and straightforward online - at one of world's fastest growing online book stores! Environmentally sound due to Print-on-Demand technologies.

Buy your books online at
www.morebooks.shop

Kaufen Sie Ihre Bücher schnell und unkompliziert online – auf einer der am schnellsten wachsenden Buchhandelsplattformen weltweit! Dank Print-On-Demand umwelt- und ressourcenschonend produziert.

Bücher schneller online kaufen
www.morebooks.shop

 info@omniscriptum.com
www.omniscriptum.com

MIX
Papier aus verantwortungsvollen Quellen
Paper from responsible sources
FSC® C105338

Printed by Books on Demand GmbH, Norderstedt / Germany